Lecture Notes in Computer Science   14631

Founding Editors

Gerhard Goos
Juris Hartmanis

The series Lecture Notes in Computer Science (LNCS), including its subseries Lecture Notes in Artificial Intelligence (LNAI) and Lecture Notes in Bioinformatics (LNBI), has established itself as a medium for the publication of new developments in computer science and information technology research, teaching, and education.

LNCS enjoys close cooperation with the computer science R & D community, the series counts many renowned academics among its volume editors and paper authors, and collaborates with prestigious societies. Its mission is to serve this international community by providing an invaluable service, mainly focused on the publication of conference and workshop proceedings and postproceedings. LNCS commenced publication in 1973.

Mario Giacobini · Bing Xue · Luca Manzoni
Editors

Genetic Programming

27th European Conference, EuroGP 2024
Held as Part of EvoStar 2024
Aberystwyth, UK, April 3–5, 2024
Proceedings

 Springer

Editors
Mario Giacobini ⓘD
University of Torino
Grugliasco, Italy

Bing Xue ⓘD
Victoria University of Wellington
Wellington, New Zealand

Luca Manzoni ⓘD
University of Trieste
Trieste, Italy

ISSN 0302-9743 ISSN 1611-3349 (electronic)
Lecture Notes in Computer Science
ISBN 978-3-031-56956-2 ISBN 978-3-031-56957-9 (eBook)
https://doi.org/10.1007/978-3-031-56957-9

This Springer imprint is published by the registered company Springer Nature Switzerland AG
The registered company address is: Gewerbestrasse 11, 6330 Cham, Switzerland

Paper in this product is recyclable.

Preface

This volume contains the proceedings of EuroGP 2024, the 27th European Conference on Genetic Programming. The conference is part of Evo*, the leading event on bio-inspired computation in Europe, and was held in Aberystwyth, Wales, UK, as a hybrid event, between Wednesday, April 3, and Friday, April 5, 2024.

EuroGP is the premier annual conference on Genetic Programming (GP), the oldest and the only meeting worldwide devoted specifically to this branch of Evolutionary Computation. At the same time, under the Evo* umbrella, EvoAPPS focused on the applications of Evolutionary Computations, EvoCOP targeted Evolutionary Computation in combinatorial optimization, and EvoMUSART was dedicated to evolved and bio-inspired music, sound, art, and design. The proceedings for these co-located events are available in the LNCS series.

Genetic Programming (GP) is a unique branch of Evolutionary Computation that has to automatically solve design problems, in particular computer program design, without requiring the user to know or specify the form or structure of the solution in advance. It uses the principles of Darwinian evolution to approach problems in the synthesis, improvement, and repair of computer programs. The universality of computer programs, and their importance in so many areas of our lives, means that the automation of these tasks is an exceptionally ambitious challenge with far-reaching implications. It has attracted a very large number of researchers and a vast amount of theoretical and practical contributions are available by consulting the GP bibliography[1].

Since the first EuroGP event in Paris in 1998, EuroGP has been the only conference exclusively devoted to the evolutionary design of computer programs and other computational structures. In fact, EuroGP represents one of the largest venues at which GP researchers meet. It plays an important role in the success of the field, by serving as a forum for expressing new ideas, meeting fellow researchers, and initiating collaborations. It attracts scholars from all over the world. In a friendly and welcoming atmosphere authors present the latest advances in the field, also presenting GP-based solutions to complex real-world problems.

EuroGP 2024 received 24 submissions from around the world. The articles underwent a rigorous double-blind peer review process, each being reviewed by by at least three Program Committee members and a senior Program Committee member.

We selected 9 of these papers for full oral presentation, while 4 works were presented in short oral presentations and as posters. In 2024, papers submitted to EuroGP could also be assigned to the "Evolutionary Machine Learning Track", with one of them accepted for oral presentation. Authors of both categories of papers also had the opportunity to present their work in poster sessions to promote the exchange of ideas in a carefree manner. All accepted contributions, regardless of the presentation format, appear as full papers in this volume.

---

[1] http://gpbib.cs.ucl.ac.uk.

An event of this kind would not be possible without the contribution of a large number of people:

– We express our gratitude to the authors for submitting their works and to the members of the Program Committee for devoting selfless effort to the review process.
– We would also like to thank Nuno Lourenço (University of Coimbra, Portugal) for his dedicated work as Submission System Coordinator.
– We thank the Evo* Graphic Identity Team, Sérgio Rebelo, Jéssica Parente and João Correia (University of Coimbra, Portugal) for their dedication and excellence in graphic design.
– We are grateful for Zakaria Abdelmoiz (University of Málaga, Spain) and João Correia (University of Coimbra, Portugal) for their impressive work managing and maintaining the Evo* website and handling the publicity, respectively.
– We credit the invited keynote speakers, Jon Timmis (Aberystwyth University, UK) and Sabine Hauert (University of Bristol, UK), for their fascinating and inspiring presentations.
– We would like to express our gratitude to the Steering Committee of EuroGP for helping organize the conference.
– Special thanks to Christine Zarges (Aberystwyth University, Wales, UK) as local organizer and to the Aberystwyth University, Wales, for organizing and providing an enriching conference venue.
– We are grateful to the support provided by SPECIES, the Society for the Promotion of Evolutionary Computation in Europe and its Surroundings, for the coordination and financial administration.

Finally, we express our continued appreciation to Anna I. Esparcia-Alcázar, from SPECIES, Europe, whose considerable efforts in managing and coordinating Evo* helped build a unique, vibrant, and friendly atmosphere.

April 2024

Mario Giacobini
Bing Xue
Luca Manzoni

Organization

Program Chairs

Mario Giacobini   University of Torino, Italy
Bing Xue   Victoria University of Wellington, New Zealand

Publication Chair

Luca Manzoni   University of Trieste, Italy

Local Chair

Christine Zarges   Aberystwyth University, Wales, UK

Publicity Chair

João Correia   University of Coimbra, Portugal

Conference Administration

Anna Esparcia-Alcázar   Evostar Coordinator

Program Committee

Wolfgang Banzhaf   Michigan State University, USA
Heder Bernardino   Universidade Federal de Juiz de Fora, Brazil
Stefano Cagnoni   University of Parma, Italy
Mauro Castelli   Universidade Nova de Lisboa, Portugal
Qi Chen   Victoria University of Wellington, New Zealand
Ernesto Costa   University of Coimbra, Portugal
Antonio Della Cioppa   University of Salerno, Italy
Steven Gustafson   Noonum, Inc., USA
Jin-Kao Hao   University of Angers, France

Contents

Short Presentations

Long Presentations

Long Presentations

Fuzzy Pattern Trees for Classification Problems Using Genetic Programming

Allan de Lima[1]([✉])[iD], Samuel Carvalho[2][iD], Douglas Mota Dias[1][iD],
Jorge Amaral[3][iD], Joseph P. Sullivan[2][iD], and Conor Ryan[1]([✉])[iD]

[1] University of Limerick, Limerick, Ireland
{Allan.Delima,Conor.Ryan}@ul.ie
[2] Technological University of the Shannon: Midlands Midwest, Limerick, Ireland
[3] Rio de Janeiro State University, Rio de Janeiro, Brazil

Abstract. Fuzzy Pattern Trees (FPTs) are tree-based structures in which the internal nodes are fuzzy operators, and the leaves are fuzzy features. This work uses Genetic Programming (GP) to evolve FPTs and assesses their performance on 20 benchmark classification problems. The results show improved accuracy for most of the problems in comparison with previous works using different approaches. Furthermore, we experiment using Lexicase Selection with FPTs and demonstrate that selection methods based on aggregate fitness, such as Tournament Selection, produce more accurate models before analysing why this is the case. We also propose new parsimony pressure methods embedded in Lexicase Selection, and analyse their ability to reduce the size of the solutions. The results show that for most problems, at least one method could reduce the size significantly while keeping a similar accuracy. We also introduce a new fuzzification scheme for categorical features with too many categories by using target encoding followed by the same scheme for numerical features, which is straightforward to implement, and avoids a much higher increase in the number of fuzzy features.

Keywords: Fuzzy Pattern Trees · Genetic Programming · Bloat control · Lexicase Selection

1 Introduction

As Artificial Intelligence (AI) continues to gain popularity by introducing many applications poised to become indispensable for humankind, the need to include explainable or interpretable traits in AI-based models has also grown. This is particularly crucial in sensitive fields such as healthcare, law and manufacturing. In addition, AI models' explanations can expand human experts' knowledge with completely new insights [21].

Fuzzy set theory can combine fuzzy terms and input features to separate the original set of features, representing them in a more comprehensible form [3]. Consequently, it can contribute to developing more interpretable solutions. Fuzzy

M. Giacobini et al. (Eds.): EuroGP 2024, LNCS 14631, pp. 3–20, 2024.
https://doi.org/10.1007/978-3-031-56957-9_1

Pattern Trees (FPTs) extend those ideas for tree-based structures, where fuzzy operators are internal nodes, while fuzzy features are leaves [15, 38]. The interpretability of FPTs was confirmed by healthcare experts in works using a heart disease dataset [27] and a sarcoidosis dataset [5].

Given the symbolic structure of GP, the solutions it evolves inherently possess a degree of interpretability, which can be further enhanced by factors such as the length of the expressions, the simplicity of the operators involved, the representation of the features, etc. For datasets of multiple types, Strongly Typed GP [23] improved the interpretability of the solutions by imposing structural constraints that prevent nonsensical operations, for instance, arithmetic operations between Boolean features. The evolution of FPTs facilitates this process because all features are represented as fuzzy features and can be used with the same sort of operators.

The work in [18] used Multi-Gene GP [9] to generate Fuzzy Inference Systems, but the first work [6] to go beyond traditional fuzzy-based rules used Cartesian GP [22] to evolve actual FPTs. Subsequently, that work was extended using a multi-objective approach, with accuracy and interpretability as objectives [29], highlighting the potential use of FPTs to bring more interpretability to the solutions. Later, Grammatical Evolution (GE) [30] was successfully used to evolve FPTs in a series of works [25–28], improving the accuracy over previous works using CGP, and contributing to the interpretability by including human knowledge [27,28]. Koza-style GP has also been used to evolve FPTs [5]. However, that work restricted their analysis to a single dataset and also represented their solutions with a single FPT using a threshold of 0.5 to discriminate between classes and, therefore, can only be used for binary classification problems.

In this work, we use GP in a generalised approach that can be applied to multi-class classification problems. Instead of using Multi-Gene GP to represent solutions as a set of tree structures, we adopt the method used in GE-based FPT research. This method represents solutions with a single tree where the root node determines the predicted class based on the highest value among its branches, with each branch corresponding to a different class. Furthermore, to improve the interpretability of FPTs, we evaluate different parsimony pressure methods to reduce the size of the solutions.

2   Background

2.1   Fuzzy Pattern Trees

FPTs are tree-like structures where internal nodes are logical and arithmetic operators constrained to yield results between 0 and 1. The leaves consist of fuzzy features, which are input features associated with fuzzy terms, valued according to the degree of membership of the input in each fuzzy set, in a process known as fuzzification [15,38]. Initially, FPTs were induced using a bottom-up approach, but later, a top-down algorithm was proposed [32]. This technique was successfully applied to classification [33] and regression problems [31].

Each FPT can be understood as a class description, which intrinsically contributes to interpretability. As a result, a solution consists of an ensemble of FPTs. In this way, when using tree-based algorithms to evolve FPTs, one could consider using multi-tree algorithms. However, the first GE-based work on FPTs [25] introduced a more straightforward approach, where the root node of an FPT was defined as Winner Takes All (WTA). This node has the same number of branches as classes in the respective problem, each branch representing an FPT itself. It predicts the class according to the highest score FPT, performing, therefore, the same action as an ensemble of FPTs. It is worth highlighting this root node is always ignored for genetic operations such as mutation.

Figure 1 presents an example of an FPT, where the input features include Proline, Color-intensity, Hue, Malic-acid, and Flavanoids. The unary operator concentrator returns the square of its operand. Since the domain of fuzzy values is between 0 and 1, this operation will reduce the value, which justifies its name. On the other hand, the operator OWA (Ordered Weighted Average) has two operands and also a weight and calculates the average between its operands, giving the weight to the biggest operand and the complement of the weight to the smallest operand. Each feature is associated with a fuzzy term that represents a specific range within the domain of that feature. For example, Proline_Low will have a high value if the value of the input feature is low. In this case, we are using the terms 'low', 'moderate' and 'high', but we could use more specific terms, such as 'cold', 'moderate' and 'hot' for an input feature related to temperature, for example. This FPT models a solution for the wine problem, which consists of classifying the samples into three types of wine, labelled as 'A', 'B', and 'C'; the expression in the left branch of the node WTA3 (Winner Takes All with three branches) represents the classification to class A, the central branch to class B, and the right branch to class C. The first branch uses only the feature Proline_Low to classify class A, and it will result in a low value if Proline_Low is high and a medium value if it is low because its complement is concentrated. In this way, we could say that if the input feature Proline presents a low value, the class is definitely not A. However, if it presents high values, the discrimination to class A will depend on the values of other branches being smaller than the

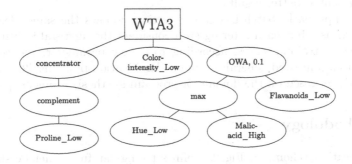

Fig. 1. Example of an FPT.

first. The second branch is easier to understand because it has no operations and uses the feature Color-intensity_Low to classify class B. In this way, if the input feature Color-intensity presents low values, the sample is more likely to be class B. On the other hand, the third branch uses three features to classify class C: Hue_Low, Malic-acid_High and Flavanoids_Low, and it will result in a high value if Flavanoids_Low is high and at least one of the others is also high. This is because the node OWA will calculate the average, giving weight 0.1 to the maximum branch and 0.9 to the minimum branch, resulting in a high value only if both branches have high values.

2.2 Lexicase Selection

Lexicase Selection is a parent selection method, originally developed to solve modal [35] and uncompromising problems [13], which has been successfully applied to different domains [2,19,24]. This method selects parents based on the performance of individuals on random permutations of the training cases taken separately instead of on an aggregate fitness score. Briefly, each selection process starts by initialising a pool of candidates with the entire population of individuals. The training cases are considered in random order, and in each step, only the candidates with the best score for the current training case are maintained in the pool. This process continues until a single candidate remains in the pool, which is then selected. Since only the best solutions for the considered training case are kept in the pool, the process can be viewed as the selection of specialists [12]. Furthermore, specialists on distinct subsets of training cases are selected as offspring since the ordering of training cases differs for each selection process. Consequently, this approach keeps high diversity [11] and promotes a better exploration of the search space [14].

Lexicase performs very well in discrete error domains, but its selection pressure is too high when employed in continuous domains, such as for regression problems, since a single training case is likely to filter most of the candidates. To overcome this, ϵ-lexicase [19] redefined the pass condition for candidates on each filtering step by defining a threshold, ϵ, as a new parameter. Nonetheless, since its original approach, an automatic ϵ, notably the Median Absolute Deviation (MAD), presented better results.

Another approach, batch-Lexicase Selection, follows the same ideas as the original Lexicase, but each filtering step considers the aggregated fitness score on a subset of training cases, according to a new parameter, the *batch size*. This approach improved generalisation on classification problems [2], and it was subsequently assessed on the domain of program synthesis problems [34].

3 Methodology

The fuzzification scheme in Fig. 2a defines triangular fuzzy membership functions that were used to fuzzify each numerical feature into three fuzzy features. Considering, for example, the input feature X, the respective fuzzy features are

X_low, X_medium and X_high. Given the value X_{example}, the respective pertinence values are μ_{low} in X_low, μ_{medium} in X_medium and 0 in X_high. On the other hand, for categorical features, we used the singleton membership function, as defined in Eq. 1 and Fig. 2b. Here, x is a categorical feature, cat_i is one of the N possible categories of that feature, and $\mu_i(x)$ is the pertinence value of a sample in a fuzzy feature that we refer to as x_i. This fuzzification scheme is analogous to one-hot encoding for categorical features.

$$\mu_i(x) \equiv \begin{cases} 1 & \text{if } x = \text{cat}_i \\ 0 & \text{otherwise} \end{cases} \tag{1}$$

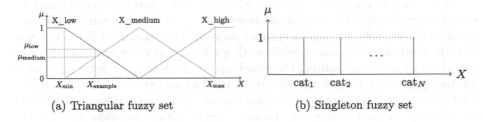

(a) Triangular fuzzy set   (b) Singleton fuzzy set

Fig. 2. Fuzzification schemes

Similarly, we use one-hot encoding to convert the target outputs into a fuzzy format, in which each sample is represented by a number of fuzzy outputs corresponding to the number of classes in the given problem. For instance, in a problem with three classes – 'A', 'B' and 'C' – 'A' is represented with 100, 'B' with 010 and 'C' with 001, each bit being considered a fuzzy output. In this example, the root node WTA3 predicts the class 'A', 'B' or 'C' based on the largest output value from the first, second, or third branch, respectively. However, we use Root Mean Squared Error (RMSE) as a fitness function between the expected fuzzy outputs and the actual values coming from each branch. This enhances the fitness score with additional information, providing more guidance during the evolutionary process. For our previous example, consider a sample with output 'A'. An individual with branch values equal to 0.6, 0.55 and 0.5 will predict class 'A' correctly, as would another with branch values equal to 0.9, 0.2 and 0.1. Still, the latter solution is preferable, and RMSE will reflect that. Equation 2 defines RMSE as we use in our work, where BV_{ij} is the branch value evaluated in the j-th branch for the i-th sample, while FO_{ij} is the fuzzy output referred to the j-th class (encoded using one-hot) for the i-th sample.

$$\text{RMSE} = \sqrt{\frac{1}{n_{\text{samples}}} \sum_{i=1}^{n_{\text{samples}}} \left(\frac{1}{n_{\text{samples}}} \sum_{j=1}^{n_{\text{samples}}} (\text{BV}_{ij} - \text{FO}_{ij})^2 \right)} \tag{2}$$

We use Eq. 2 to calculate an aggregated fitness score from all samples, which is then used with tournament selection. However, since we also run experiments with Lexicase-based approaches, we need to define a fitness case function. We tried some preliminary experiments using the equivalent RMSE as a fitness case function, and noticed the evolution was quite poor. This happened because the distance between small errors is bigger using RMSE than MSE, and since Lexicase considers the fitness cases separately, it increases the pressure too much in the filtering steps. Equation 3 defines MSE as we use in our work, where, for a given sample, BV_i is the branch value evaluated in the i-th branch, while FO_i is the fuzzy output referred to the i-th class.

$$\text{MSE} = \frac{1}{n_{\text{classes}}} \sum_{i=1}^{n_{\text{classes}}} (BV_i - FO_i)^2 \tag{3}$$

In addition to promoting accuracy, we also wish to encourage small solutions to aid interpretation. GP implementations usually use the maximum depth as a parameter. However, setting an optimal value for this parameter is too difficult since it is problem-dependent, while setting a too-small value could hamper the evolution by over-constraining the search space. Therefore, we set a high value for this parameter to give the algorithm freedom to explore the search space. Furthermore, we aim to analyse the performance of different parsimony methods in reducing size and setting the maximum depth with a low value to force the creation of small individuals would compromise our analysis.

The most common method to add parsimony pressure to the selection process involves penalising solutions in proportion to their size, introducing a penalty ratio as a new parameter. Another option is lexicographic parsimony pressure [20], which consists of preferring smaller individuals when there is a tie in the fitness score. Lexi[2] [4] applies this approach to Lexicase Selection [35]. Given the nature of our fitness evaluations, it is anticipated that each individual will possess unique fitness case scores. Consequently, employing Lexicase Selection could allow the selection of a parent based on a single sample, resulting in excessively high selection pressure. Thus, using epsilon-lexicase, which introduces a threshold of tolerance and applying lexicographic parsimony pressure to epsilon-lexicase, implementing epsilon-lexi[2] appears to be a viable alternative, even though we are tackling classification problems.

Following the recommendations of the original work for epsilon-lexicase, we use MAD as epsilon for epsilon-lexi[2]. In addition, the original work also recommends calculating MAD once per generation instead of at each filtering step because the difference in the results is not significant, and it saves computational costs. In this way, MAD is computed for each sample across all individuals at the outset. Subsequently, imagine we have a vector for each sample containing the fitness case score for every individual with respect to that sample. We identify the minimum value of that vector as the best fitness case score for that sample. Given the corresponding MAD, we can replace a value in that vector with 0 if it is smaller or equal to (best + MAD) and with 1 otherwise.

After processing each sample in this manner, we obtain a vector of discrete fitness cases to represent each individual. Then, we initialise the pool of candidates using only individuals with unique error vectors. This prefiltering step reduces the time spent in the selection process by avoiding redundant iterations when filtering the pool of candidates, and it does not affect the outcome [10, 12]. During this step, we also apply lexicographic parsimony pressure by selecting the smallest individual in cases where multiple individuals share an identical error vector. In terms of the individual selected, this has the same effect as if we have not prefiltered the pool and chosen the smallest individual after filtering the pool with all fitness cases.

---

1. **Initialise:**
   (a) Place all individuals in a pool of `candidates`
   (b) Create batches by taking different training cases in random order and list them in `batch_cases`. All batches have `batch_size` cases, except the last one, which has the remaining cases
   (c) Calculate the aggregated fitness for each batch by averaging the fitness values of the respective samples
   (d) For each batch, calculate MAD and identify the best fitness over all individuals
   (e) For each batch in each individual, change the value to 0 if it is smaller or equal to the respective best value plus the respective MAD and to 1 otherwise, in order to create for each individual an error vector with discrete values
   (f) Keep in the pool of `candidates` only individuals with unique error vectors
      i. When individuals with the same error vector are found, keep the one with the smallest number of nodes
      ii. If there is a tie within the number of nodes, keep a random individual
2. **Loop:**
   (a) Replace `candidates` with the individuals currently in `candidates`, which presented the best fitness for the first batch in `batch_cases`
   (b) If a single individual remains in `candidates`, return this individual
   (c) Else eliminate the first element in `batch_cases` and re-run the `Loop`

Listing 1. Algorithm for selecting one individual with batch-epsilon-lexi[2].

In addition, we also used the epsilon as a threshold and applied lexicographic parsimony pressure to an approach based on batch-Lexicase, which we refer to as batch-epsilon-lexi[2]. Listing 1 details the algorithm for selecting one individual with this approach. We begin by populating the batches with randomly selected cases. Next, we calculate the fitness batch scores based on adding the respective fitness case scores for each individual. Subsequently, we calculate the MAD and identify the best score for each batches. After that, we can replace the continuous values with 0 or 1 and prefilter individuals with the same error vectors, preferring the smallest ones. Finally, the `Loop` stage is similar to standard Lexicase Selection. This approach is more expensive than epsilon-lexi[2], because we need to calculate MAD every time for the event of selecting a parent instead of once per generation. On the other hand, the computational cost in the filtering process is reduced because the bigger the batch size, the smaller the number of filtering steps, as we show in our results. However, overall, this approach is still more expensive.

In summary, the scenarios we analyse in this work are as follows:

- tournament: this is our baseline approach, where we use tournament selection to select parents based on the RMSE;
- tournament with penalised fitness: in this scenario, we also use tournament selection, but the fitness score is the RMSE plus the number of nodes divided by 500. We use the number of nodes as size measurement as this gives more diverse values than depth. In relation to the penalty factor, we experimented with various values. Preliminary results indicated that a factor of 500 effectively reduces the size of the solutions without adversely affecting the error rate;
- epsilon-lexi^2: we use epsilon-lexicase to select parents based on the MSE for each sample and apply lexicographic parsimony pressure by using the number of nodes as a tie-breaking criterion to prefer the smallest individual;
- batch-epsilon-lexi^2 (batch size 2): we use batch-epsilon-lexicase to select parents based on the MSE for each batch of samples in this and in the remaining scenarios. We performed experiments in a range of small batch sizes, because for bigger sizes, the selection pressure is usually much smaller [34]. In addition, we apply lexicographic parsimony pressure by using the number of nodes as a tie-breaking criterion to prefer the smallest individual;
- batch-epsilon-lexi^2 (batch size 5);
- batch-epsilon-lexi^2 (batch size 10);
- batch-epsilon-lexi^2 (batch size 20).

4   Experimental Setup

We used Python 3.10.8 and DEAP 1.4 [8] to run our experiments, seeding each run with `random.seed(n)`, where n is an integer number in the interval [1, 30], to replicate our results. All code is available in our repository[1].

Table 1 provides a summary of the classification problems examined in this work. Some of them are "blacklisted" [36], but we still assessed the performance of our implementation with these datasets to compare our results with previous works using FPTs. In addition, we included some datasets from the CHIRP suite [37], following the recommendation of using it as an alternative to the "blacklisted" ploblems [36]. Consequently, our selection encompasses a diverse array of problems, varying in the number of classes, class balance, sample size, feature types, and so on. All datasets are accessible via the links provided in the references. In instances where the data were pre-partitioned into training and test sets, we adhered to this division and have indicated the respective sizes in the table. Conversely, for datasets that were not pre-divided, we used 75% for training and 25% for testing, performing a fresh split for each run. For these cases, the table displays the total size of the dataset. Additionally, we have detailed the number of numerical and categorical features, as well as the total number of features after fuzzifying the data.

---

[1] https://github.com/bdsul/grape/tree/main/GP/ClassificationFPT

Table 1. Characteristics of datasets, taken from the UCI repository [7], the CMU repository [17] and the ProPublica data store [1].

	Classes	Training	Testing	Numerical features	Categorical features	Features after Fuzzification
australian	2	688		8	6	32
adult	2	32561	16281	5	9	40
bankMarketing	2	41188		9	11	58
credit	2	690		6	9	37
germanCredit	2	1000		7	13	73
haberman	2	306		3	–	9
heartDisease	2	297		5	8	35
horse	3	300	68	2	18	58
iris	3	150		4	–	12
lawsuit	2	264		3	1	10
lupus	2	87		3	–	9
pima	2	768		8	–	24
recidivism	2	6172		4	5	25
satellite	6	4435	2000	36	–	108
segment	7	210	2100	16	2	54
spect	2	80	187	–	22	22
transfusion	2	748		4	–	12
vehicle	3	846		18	–	54
violentRecidivism	2	4743		5	4	23
wine	3	178		13	–	39

We used the schemes detailed in Fig. 2 to fuzzify features. However, for categorical features with too many different categories, that scheme would create an impractical number of fuzzy features, so we propose in this work using target encoding followed by the scheme for numerical features. All fuzzification thresholds were defined based on the training set to avoid bias towards the test set. For example, in Fig. 2a, $X_{\min}$ and $X_{\max}$ are set according to the training set. If the respective feature in the test set presents a value smaller than $X_{\min}$ or greater than $X_{\max}$, its fuzzy pertinence value will be 0 or 1, respectively. Similarly, target encoding followed only the values from the training set to encode both the training and test sets.

Table 2. Experimental setup.

(a) Model parameters.

Parameter type	Parameter value
Number of runs	30
Number of generations	50
Population size	500
Maximum depth	17
Mutation probability	0.05
Crossover probability	0.8
Initialisation method	Ramped half-and-half
Minimum initial depth	2
Maximum initial depth	6
Tournament size	7
epsilon	MAD
Lexi^2 criterion	Number of nodes

(b) Fuzzy operators used in the function set. WA means Weighted Average, while OWA means Ordered Weighted Average.

Operator	Equivalent expression
max(a, b)	–
min(a, b)	–
WA(a, b, r)	$r \times a + (1 - r) \times b$
OWA(a, b, r)	$r \times \max(a, b) +$ $+ (1 - r) \times \min(a, b)$
dilator(a)	$\sqrt{a}$
concentrator(a)	a^2
complement(a)	$1 - a$

Table 2a summarises the parameters used in our experiments, while Table 2b describes the meaning of the fuzzy operators used in our function set. Regarding the operators WA and OWA, the operand r is constrained to use a constant from the set $\{0.1, 0.2, 0.3, 0.4, 0.5, 0.6, 0.7, 0.8, 0.9\}$.

5   Results and Discussion

Table 3 presents the mean accuracy obtained on the test set across 30 trials of our work employing tournament selection, alongside results from other studies using CGP [6] and GE [25, 27] for the works we have problems in common, where we used the same population size and number of generations. The results, with superior accuracies denoted in bold, indicate that the GP approach yielded the highest accuracy in 10 out of 14 problems.

Table 4 provides a detailed summary of our study's outcomes, displaying for each dataset the mean accuracy on the test set alongside the average node count of the optimal individual.

We conducted a statistical analysis to evaluate the significance of these findings. The Shapiro-Wilk test was applied to both

Table 3. Mean of the accuracy computed on the test set over 30 runs using plain tournament with GP compared to the results using CGP [6] and GE [25,27], where the best results are in bold.

	GE	CGP	GP
adult	0.81	–	**0.83**
australian	**0.86**	0.85	0.85
bankMarketing	0.89	–	**0.90**
germanCredit	0.71	–	**0.72**
haberman	**0.74**	0.73	**0.74**
heartDisease	0.79	–	**0.83**
iris	**0.96**	0.95	**0.96**
lawsuit	**0.96**	0.93	0.95
lupus	0.73	0.74	**0.75**
pima	0.74	0.72	**0.76**
recidivism	**0.72**	–	0.68
transfusion	0.77	0.76	**0.78**
violentRecidivism	**0.83**	–	0.81
wine	0.83	0.9	**0.91**

Table 4. Mean (standard deviation) of the accuracy computed on the test set and the number of nodes of the best individual. Both were calculated over 30 runs with the respective method. TPP refers to tournament with parsimony pressure, while B-ϵ-L^2 refers to batch-epsilon-lexi^2, where the number between parenthesis is the batch size. The first row for each dataset refers to the accuracy, and the second row refers to the size. The symbols $+$, $-$, $=$ indicate, for each method other than tournament, whether the results in comparison to tournament are, respectively, significantly better, worse, or not significantly different. The last row indicates how many datasets the respective method presented results, which are not significantly worse than tournament in accuracy and significantly better in size.

	tournament	TPP	ϵ-lexi^2	B-ϵ-L^2(2)	B-ϵ-L^2(5)	B-ϵ-L^2(10)	B-ϵ-L^2(20)
adult	0.83(0.01)	0.82(0.01)−	0.82(0.00)−	0.82(0.00)−	0.83(0.00)=	0.83(0.00)−	0.83(0.00)=
	100.8(44.3)	55.0(32.9)+	39.2(28.8)+	35.2(12.7)+	51.0(25.8)+	61.6(33.3)+	87.1(33.9)=
australian	0.85(0.02)	0.85(0.02)=	0.85(0.02)=	0.86(0.02)=	0.86(0.02)=	0.86(0.02)=	0.86(0.02)=
	120.0(67.1)	59.4(32.5)+	72.3(34.1)+	53.0(22.8)+	76.2(24.3)+	86.3(23.9)+	108.9(38.3)=
bankMarketing	0.90(0.00)	0.90(0.00)−	0.89(0.00)−	0.89(0.00)−	0.90(0.00)=	0.90(0.00)=	0.90(0.00)=
	72.2(36.4)	39.2(21.5)+	18.8(11.0)+	17.6(9.9)+	28.2(12.9)+	48.3(18.5)+	58.3(22.0)=
credit	0.86(0.03)	0.86(0.03)=	0.86(0.03)=	0.85(0.03)=	0.86(0.03)=	0.85(0.03)=	0.86(0.02)=
	96.5(48.1)	67.9(35.3)=	56.5(24.8)+	53.0(23.8)+	72.6(22.3)=	79.4(30.4)=	95.2(36.0)=
germanCredit	0.72(0.03)	0.70(0.03)=	0.70(0.03)=	0.70(0.03)=	0.71(0.03)=	0.72(0.03)=	0.72(0.03)=
	131.4(46.4)	97.8(34.0)+	45.5(22.8)+	42.3(17.1)+	58.9(35.2)+	85.1(39.8)+	113.6(38.6)=
haberman	0.74(0.05)	0.74(0.05)=	0.74(0.05)=	0.74(0.05)=	0.74(0.06)=	0.73(0.06)=	0.74(0.05)=
	139.8(49.1)	73.6(40.6)+	36.3(18.0)+	36.9(22.9)+	51.0(20.2)+	74.7(27.3)+	107.0(39.0)+
heartDisease	0.83(0.04)	0.81(0.04)=	0.79(0.05)−	0.81(0.04)=	0.81(0.04)=	0.82(0.03)=	0.82(0.04)=
	129.4(42.7)	88.8(36.7)+	48.8(23.6)+	72.5(30.1)+	99.8(35.8)+	102.4(34.9)=	125.0(46.6)=
horse	0.73(0.04)	0.72(0.03)=	0.70(0.03)=	0.72(0.04)=	0.72(0.02)=	0.74(0.04)=	0.73(0.05)=
	107.0(51.7)	80.3(42.5)=	35.0(14.5)+	40.3(16.7)+	45.7(17.2)+	74.9(33.5)+	93.4(34.1)=
iris	0.96(0.03)	0.95(0.03)=	0.95(0.03)=	0.96(0.03)=	0.95(0.03)=	0.95(0.03)=	0.95(0.03)=
	60.8(67.5)	14.1(5.4)+	21.0(16.3)+	26.9(19.3)+	33.4(19.4)+	27.0(17.8)+	23.7(24.3)+
lawsuit	0.95(0.02)	0.96(0.02)=	0.95(0.03)=	0.95(0.02)=	0.95(0.02)=	0.95(0.02)=	0.96(0.03)=
	109.1(56.1)	48.8(31.1)+	21.5(16.3)+	22.5(12.4)+	36.3(26.5)+	53.8(31.7)+	77.4(26.4)+
lupus	0.75(0.07)	0.74(0.07)=	0.74(0.06)=	0.74(0.07)=	0.73(0.07)=	0.74(0.08)=	0.74(0.08)=
	79.2(60.3)	47.0(30.4)+	39.8(24.4)+	38.8(27.8)+	59.4(26.0)=	72.1(24.5)=	94.6(43.4)=
pima	0.76(0.02)	0.75(0.02)=	0.76(0.03)=	0.75(0.02)=	0.76(0.03)=	0.76(0.02)−	0.76(0.03)−
	99.7(60.2)	49.7(32.0)+	26.6(10.9)+	37.2(22.3)+	41.2(17.2)+	57.5(22.0)+	74.3(18.2)=
recidivism	0.68(0.01)	0.67(0.01)=	0.66(0.01)−	0.67(0.01)−	0.67(0.01)−	0.68(0.01)=	0.68(0.01)=
	102.7(48.5)	90.6(38.7)=	43.1(24.9)+	33.1(16.3)+	48.9(26.2)+	57.9(23.6)+	72.1(27.3)+
satellite	0.69(0.03)	0.66(0.03)−	0.57(0.06)−	0.69(0.03)=	0.70(0.02)=	0.72(0.02)+	0.72(0.04)+
	48.3(26.3)	33.7(15.1)=	18.9(7.8)+	36.8(14.1)=	45.0(16.9)=	52.9(15.7)=	59.8(31.7)=
segment	0.73(0.05)	0.61(0.08)−	0.48(0.08)−	0.73(0.05)=	0.77(0.04)+	0.76(0.08)=	0.76(0.09)=
	44.3(16.6)	34.1(14.4)=	23.2(6.7)+	34.3(9.6)+	39.6(13.7)=	49.0(17.7)=	59.5(22.8)−
spect	0.77(0.04)	0.75(0.05)=	0.78(0.04)=	0.78(0.03)=	0.78(0.03)=	0.77(0.03)=	0.76(0.04)=
	165.3(46.5)	125.6(45.8)+	75.4(25.1)+	84.1(28.2)+	114.8(40.7)+	120.5(29.7)+	160.0(54.5)=
transfusion	0.78(0.03)	0.77(0.03)=	0.77(0.03)=	0.77(0.03)=	0.78(0.03)=	0.78(0.03)=	0.78(0.03)=
	121.3(58.6)	73.5(44.4)+	40.3(17.1)+	40.0(18.8)+	64.4(27.0)+	77.4(35.9)+	96.3(40.5)=
vehicle	0.69(0.05)	0.66(0.04)−	0.64(0.04)−	0.69(0.04)=	0.72(0.04)+	0.72(0.04)=	0.73(0.04)+
	59.3(30.1)	37.8(31.1)+	13.0(8.9)+	34.3(13.6)+	52.4(24.2)=	64.0(22.7)=	83.1(32.8)−
violentRecidivism	0.81(0.01)	0.80(0.01)−	0.80(0.01)=	0.80(0.01)=	0.80(0.01)=	0.80(0.01)=	0.81(0.01)=
	121.9(57.9)	76.3(46.6)+	31.6(26.9)+	25.2(14.7)+	46.1(22.1)+	53.7(27.3)+	69.2(28.4)+
wine	0.91(0.04)	0.91(0.05)=	0.90(0.05)=	0.91(0.05)=	0.93(0.04)=	0.92(0.05)=	0.92(0.04)=
	60.9(35.2)	24.7(15.3)+	40.8(16.2)+	55.0(18.9)=	60.1(24.6)=	64.8(31.8)=	57.5(37.0)=
Total		12	13	15	12	11	5

the accuracy and the number of nodes to test for normality, with a p-value threshold set at 0.05. To compare the performance of various approaches against the baseline employing tournament selection, we utilised the Wilcoxon signed-rank test for the non-Gaussian results, and the Student's t-test for the Gaussian results. However, to account for the multiple comparisons being made, we applied a Bonferroni correction, adjusting the significance level by a factor of 6. Consequently, we adopted a more stringent p-value threshold of 0.0083, resulting from the division of 0.05 by 6, to reject the null hypothesis (no difference between the metrics).

To conserve space, we have omitted the explicit reporting of calculated p-values. Instead, we employed the symbols $+$, $-$, $=$ to denote whether the results, in comparison to tournament selection, are significantly better, worse, or not significantly different, respectively. Consider that for accuracy, being better means it increased, while for size, being better means it decreased. We noted that, with the exception of the satellite dataset, every problem exhibited at least one approach in which accuracy was maintained without significant reduction, while the application of parsimony pressure methods resulted in a substantial decrease in size.

In addition, the final row of the table indicates the number of datasets for which each method yielded results that were not significantly inferior to tournament selection in terms of accuracy yet were significantly more efficient in terms of size reduction. We observed the most efficient method for reducing size without compromising accuracy was batch-epsilon-lexi[2] using a small batch size (2). Furthermore, this same method, utilising larger batch sizes (5, 10 and 20), was able to significantly enhance accuracy compared to tournament selection on a total of five occasions. Nonetheless, ultimately, the approach using tournament still provided the best accuracy for most of the problems. As discussed in Sect. 3, defining the aggregated fitness as the RMSE between the values from the ensemble of FPTs and the one-hot representation of the target outputs brings much information to the aggregated score, making tournament selection a highly competitive selection method.

Figure 3 presents some generational results for other measurements using the heart disease dataset, but similar observations can be made for other problems. All graphs are plotted over an average of 30 runs. Figure 3a shows the average MAD across generations. Firstly, MAD is taken for each sample across all individuals, and then this graph illustrates the average MAD over all samples. We also report these values for tournament-based approaches, even though it is not being used for those methods, in order to clarify what is happening during the evolution for all scenarios. Figure 3b shows the average variance of the fitness case values across generations. For this graph, we calculate the variance of the MSE values for every sample for all individuals. This measurement is not considered in the evolutionary process but is good for analysing the spread of fitness case values. Figure 3c depicts the average number of filtering steps with the Lexicase-based methods across generations. This measurement refers to the number of fitness case values used by the Lexicase loop to filter the pool of candidates,

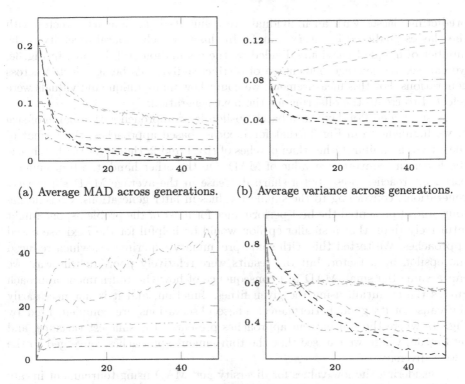

(a) Average MAD across generations.   (b) Average variance across generations.

(c) Average number of filtering steps with (d) Average behavioural diversity of the
Lexicase-based methods across genera- predictions across generations.
tions.

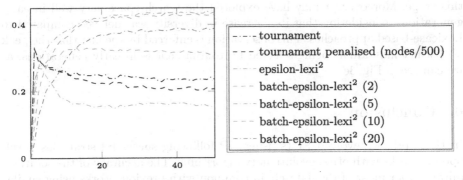

-----tournament
--- tournament penalised (nodes/500)
--- epsilon-lexi^2
--- batch-epsilon-lexi^2 (2)
--- batch-epsilon-lexi^2 (5)
--- batch-epsilon-lexi^2 (10)
-----batch-epsilon-lexi^2 (20)

(e) Average percentage of distinct individuals being selected across generations.

Fig. 3. Generational results for some measurements with the heart disease dataset.

and it is used to confirm an assumption we made in Sect. 3, where we stated the
bigger the batch size, the smaller the number of filtering steps. In Fig. 3d, we can
see the average behavioural diversity of the predictions across generations. After

predicting classes with an individual, we define its behaviour as a vector with the classes predicted for every sample. In this way, behavioural diversity is the number of unique behaviours divided by the population [16]. Finally, in Fig. 3e, we can see the average percentage of distinct individuals being selected across generations. For this measurement, we count how many unique individuals were selected to create the offspring for the next generation.

Assuming epsilon-lexi^2 as a batch-epsilon-lexi^2 approach with a batch of size 1, we highlight from Fig. 3a that for Lexicase-based approaches, the format of the curves is similar. Other than batches of size 1 and 2, the larger the batch size the lower the convergence value of MAD. On the other hand, the tournament-based approaches presented a sharp decrease in the average MAD since early generations, converging to the smallest values in later generations. Considering tournament presented the best performance for most of the problems, one might intuitively think that a smaller epsilon would be helpful for the Lexicase-based approaches. We tested this with some preliminary experiments, which reduced the epsilon by a factor, but the results were relatively poor. In this way, we can assume the small MAD is a consequence of how the tournament approach guides the evolution using the given fitness function, and it is not necessarily the cause of its good performance. These observations are complemented by Fig. 3b, where the tournament approaches presented the smallest variance, and by Fig. 3d, where we can see that the tournament approaches converged to the lowest diversity.

Considering the low values for diversity and MAD using tournament in late generations, we can assume the individuals in these populations are more similar to each other than in the Lexicase-based approaches. In this way, since tournament exhibits the best performance, we can assume it is exploiting very well at this stage. Moreover, it may have explored the search space very well in early generations, considering that its decrease in diversity was delayed compared to Lexicase-based approaches. This observation is enforced by the fact that the peak of distinct individuals being selected by tournament is in early generations, as we can see in Fig. 3e.

6   Conclusion

In this work, we evolved FPTs using GP following successful strategies developed in works with other evolutionary algorithms. The accuracy of the solutions improved for most of the datasets in common with previous works using similar computational costs (population size and number of generations). In addition, we proposed using Lexicase Selection to evolve FPTs, as well as a new fuzzification scheme for categorical features with too many categories by using target encoding followed by a scheme for numerical features.

Moreover, we evaluated the performance of the system across a broad range of datasets and proposed changes in well-known selection methods to include parsimony pressure strategies. For the majority of the problems investigated, we identified at least one method capable of significantly diminishing the complexity of the models without compromising their performance, which remained comparable to that achieved by tournament selection.

In addition, we showed that in terms of accuracy, plain tournament and batch-epsilon-lexi[2] resulted in the best performance, highlighting that for FPTs, selection methods based on aggregated fitness scores worked best on these problems.

Although the bottom-up approach to induce FPTs was already applied for regression problems, as we highlighted in Sect. 2, there is still a lack in the literature about using evolutionary algorithms for that. In future work, we plan to extend the strategies presented here to apply GP to synthesising FPTs for regression problems.

Representing input data using fuzzy logic usually brings more interpretability to the features since it separates the data into specific, often meaningful, parts of their domain, usually associated with a descriptive term. In this way, we also intend, in future work, to deeply analyse FPTs from the perspective of interpretability, since it has become an essential concern for Machine Learning (ML) models in critical areas such as healthcare, law and manufacturing industries.

References

1. Propublica data store. https://www.propublica.org/datastore/dataset/compas-recidivism-risk-score-data-and-analysis
2. Aenugu, S., Spector, L.: Lexicase selection in learning classifier systems. In: Proceedings of the Genetic and Evolutionary Computation Conference, pp. 356–364 (2019). https://doi.org/10.1145/3321707.3321828
3. Cordón, O.: A historical review of evolutionary learning methods for Mamdani-type fuzzy rule-based systems: designing interpretable genetic fuzzy systems. Int. J. Approximate Reasoning **52**(6), 894–913 (2011). https://doi.org/10.1016/j.ijar.2011.03.004
4. de Lima, A., Carvalho, S., Dias, D.M., Naredo, E., Sullivan, J.P., Ryan, C.: Lexi2: lexicase selection with lexicographic parsimony pressure. In: Proceedings of the Genetic and Evolutionary Computation Conference, GECCO 2022, pp. 929–937. Association for Computing Machinery, New York (2022). https://doi.org/10.1145/3512290.3528803
5. de Lima, A.D., Lopes, A.J., do Amaral, J.L.M., de Melo, P.L.: Explainable machine learning methods and respiratory oscillometry for the diagnosis of respiratory abnormalities in sarcoidosis. BMC Med. Inform. Decis. Making **22**(1), 274 (2022). https://doi.org/10.1186/s12911-022-02021-2
6. Dos Santos, A.R., Amaral, J.L.M.: Synthesis of fuzzy pattern trees by cartesian genetic programming. Mathware Soft Comput.: Mag. Eur. Soc. Fuzzy Log. Technol. **22**(1), 52–56 (2015)
7. Dua, D., Graff, C.: UCI machine learning repository (2017). http://archive.ics.uci.edu/ml

8. Fortin, F.A., De Rainville, F.M., Gardner, M.A.G., Parizeau, M., Gagné, C.: DEAP: evolutionary algorithms made easy. J. Mach. Learn. Res. **13**(1), 2171–2175 (2012)
9. Gandomi, A.H., Alavi, A.H.: A new multi-gene genetic programming approach to nonlinear system modeling. Part I: materials and structural engineering problems. Neural Comput. Appl. **21**(1), 171–187 (2012). https://doi.org/10.1007/s00521-011-0734-z
10. Helmuth, T., Lengler, J., La Cava, W.: Population diversity leads to short running times of lexicase selection. In: Rudolph, G., Kononova, A.V., Aguirre, H., Kerschke, P., Ochoa, G., Tušar, T. (eds.) PPSN 2022. LNCS, vol. 13399, pp. 485–498. Springer, Cham (2022). https://doi.org/10.1007/978-3-031-14721-0_34
11. Helmuth, T., McPhee, N.F., Spector, L.: Lexicase selection for program synthesis: a diversity analysis. In: Riolo, R., Worzel, B., Kotanchek, M., Kordon, A. (eds.) Genetic Programming Theory and Practice XIII. GEC, pp. 151–167. Springer, Cham (2016). https://doi.org/10.1007/978-3-319-34223-8_9
12. Helmuth, T., Pantridge, E., Spector, L.: On the importance of specialists for lexicase selection. Genet. Program Evolvable Mach. **21**(3), 349–373 (2020). https://doi.org/10.1007/s10710-020-09377-2
13. Helmuth, T., Spector, L., Matheson, J.: Solving uncompromising problems with lexicase selection. IEEE Trans. Evol. Comput. **19**(5), 630–643 (2015). https://doi.org/10.1109/TEVC.2014.2362729
14. Hernandez, J.G., Lalejini, A., Ofria, C.: An exploration of exploration: measuring the ability of lexicase selection to find obscure pathways to optimality. In: Banzhaf, W., Trujillo, L., Winkler, S., Worzel, B. (eds.) Genetic Programming Theory and Practice XVIII. Genetic and Evolutionary Computation, pp. 83–107. Springer, Singapore (2022). https://doi.org/10.1007/978-981-16-8113-4_5
15. Huang, Z., Gedeon, T.D., Nikravesh, M.: Pattern trees induction: a new machine learning method. IEEE Trans. Fuzzy Syst. **16**(4), 958–970 (2008). https://doi.org/10.1109/TFUZZ.2008.924348
16. Jackson, D.: Promoting phenotypic diversity in genetic programming. In: Schaefer, R., Cotta, C., Kołodziej, J., Rudolph, G. (eds.) PPSN 2010. LNCS, vol. 6239, pp. 472–481. Springer, Heidelberg (2010). https://doi.org/10.1007/978-3-642-15871-1_48
17. Kooperberg, C.: StatLib: an archive for statistical software, datasets, and information. Am. Stat. **51**(1), 98 (1997)
18. Koshiyama, A.S., Vellasco, M.M.B.R., Tanscheit, R.: GPFIS-CLASS: a genetic fuzzy system based on genetic programming for classification problems. Appl. Soft Comput. **37**, 561–571 (2015). https://doi.org/10.1016/j.asoc.2015.08.055
19. La Cava, W., Spector, L., Danai, K.: Epsilon-lexicase selection for regression. In: Proceedings of the Genetic and Evolutionary Computation Conference 2016, GECCO 2016, pp. 741–748. Association for Computing Machinery, New York (2016). https://doi.org/10.1145/2908812.2908898
20. Luke, S., Panait, L.: Lexicographic parsimony pressure. In: Proceedings of the 4th Annual Conference on Genetic and Evolutionary Computation, GECCO 2002, pp. 829–836. Morgan Kaufmann Publishers Inc., San Francisco (2002)
21. Mei, Y., Chen, Q., Lensen, A., Xue, B., Zhang, M.: Explainable artificial intelligence by genetic programming: a survey. IEEE Trans. Evol. Comput. **27**(3), 621–641 (2023). https://doi.org/10.1109/TEVC.2022.3225509
22. Miller, J.F.: Cartesian genetic programming. In: Miller, J.F. (ed.) Cartesian Genetic Programming. Natural Computing Series, pp. 17–34. Springer, Heidelberg (2011). https://doi.org/10.1007/978-3-642-17310-3_2

23. Montana, D.J.: Strongly typed genetic programming. Evol. Comput. **3**(2), 199–230 (1995). https://doi.org/10.1162/evco.1995.3.2.199
24. Moore, J.M., Stanton, A.: Lexicase selection outperforms previous strategies for incremental evolution of virtual creature controllers. In: ECAL 2017, the Fourteenth European Conference on Artificial Life, pp. 290–297. MIT Press (2017). https://doi.org/10.1162/isal_a_050
25. Murphy, A., Ali, M., Dias, D., Amaral, J., Naredo, E., Ryan, C.: Grammar-based fuzzy pattern trees for classification problems. In: Proceedings of the 12th International Joint Conference on Computational Intelligence, pp. 71–80. SCITEPRESS - Science and Technology Publications, Budapest (2020). https://doi.org/10.5220/0010111900710080
26. Murphy, A., Ali, M.S., Mota Dias, D., Amaral, J., Naredo, E., Ryan, C.: Fuzzy pattern tree evolution using grammatical evolution. SN Comput. Sci. **3**(6), 426 (2022). https://doi.org/10.1007/s42979-022-01258-y
27. Murphy, A., Murphy, G., Amaral, J., MotaDias, D., Naredo, E., Ryan, C.: Towards incorporating human knowledge in fuzzy pattern tree evolution. In: Hu, T., Lourenço, N., Medvet, E. (eds.) EuroGP 2021. LNCS, vol. 12691, pp. 66–81. Springer, Cham (2021). https://doi.org/10.1007/978-3-030-72812-0_5
28. Murphy, A., Murphy, G., Dias, D.M., Amaral, J., Naredo, E., Ryan, C.: Human in the loop fuzzy pattern tree evolution. SN Comput. Sci. **3**(2), 163 (2022). https://doi.org/10.1007/s42979-022-01044-w
29. Rodrigues dos Santos, A., Machado do Amaral, J.L., Ribeiro Soares, C.A., Valladão de Barros, A.: Multi-objective fuzzy pattern trees. In: 2018 IEEE International Conference on Fuzzy Systems (FUZZ-IEEE), pp. 1–6 (2018). https://doi.org/10.1109/FUZZ-IEEE.2018.8491689
30. Ryan, C., Collins, J.J., Neill, M.O.: Grammatical evolution: evolving programs for an arbitrary language. In: Banzhaf, W., Poli, R., Schoenauer, M., Fogarty, T.C. (eds.) EuroGP 1998. LNCS, vol. 1391, pp. 83–96. Springer, Heidelberg (1998). https://doi.org/10.1007/BFb0055930
31. Senge, R., Hüllermeier, E.: Pattern trees for regression and fuzzy systems modeling. In: International Conference on Fuzzy Systems, pp. 1–7 (2010). https://doi.org/10.1109/FUZZY.2010.5584231
32. Senge, R., Hüllermeier, E.: Top-down induction of fuzzy pattern trees. IEEE Trans. Fuzzy Syst. **19**(2), 241–252 (2011). https://doi.org/10.1109/TFUZZ.2010.2093532
33. Shaker, A., Senge, R., Hüllermeier, E.: Evolving fuzzy pattern trees for binary classification on data streams. Inf. Sci. **220**, 34–45 (2013). https://doi.org/10.1016/j.ins.2012.02.034
34. Sobania, D., Rothlauf, F.: Program synthesis with genetic programming: the influence of batch sizes. In: Medvet, E., Pappa, G., Xue, B. (eds.) EuroGP 2022. LNCS, vol. 13223, pp. 118–129. Springer, Cham (2022). https://doi.org/10.1007/978-3-031-02056-8_8
35. Spector, L.: Assessment of problem modality by differential performance of lexicase selection in genetic programming: a preliminary report. In: Proceedings of the 14th Annual Conference Companion on Genetic and Evolutionary Computation, GECCO 2012, pp. 401–408. Association for Computing Machinery, New York (2012). https://doi.org/10.1145/2330784.2330846
36. White, D.R., et al.: Better GP benchmarks: community survey results and proposals. Genet. Program Evolvable Mach. **14**(1), 3–29 (2013). https://doi.org/10.1007/s10710-012-9177-2

37. Wilkinson, L., Anand, A., Tuan, D.N.: CHIRP: a new classifier based on composite hypercubes on iterated random projections. In: Proceedings of the 17th ACM SIGKDD International Conference on Knowledge Discovery and Data Mining, pp. 6–14. KDD 2011, Association for Computing Machinery, New York (2011). https://doi.org/10.1145/2020408.2020418

38. Yi, Y., Fober, T., Hüllermeier, E.: Fuzzy operator trees for modeling rating functions. Int. J. Comput. Intell. Appl. **08**(04), 413–428 (2009). https://doi.org/10.1142/S1469026809002679

Generational Computation Reduction in Informal Counterexample-Driven Genetic Programming

Thomas Helmuth[1](✉) ⓘ, Edward Pantridge[2] ⓘ, James Gunder Frazier[1] ⓘ,
and Lee Spector[3,4] ⓘ

[1] Hamilton College, Clinton, NY 13323, USA
{thelmuth,jgfrazie}@hamilton.edu
[2] Real Chemistry, Boston, MA 02111, USA
ed@swoop.com
[3] Amherst College, Amherst, MA 01002, USA
lspector@amherst.edu
[4] University of Massachusetts, Amherst, MA 01003, USA

Abstract. Counterexample-driven genetic programming (CDGP) uses specifications provided as formal constraints to generate the training cases used to evaluate evolving programs. It has also been extended to combine formal constraints and user-provided training data to solve symbolic regression problems. Here we show how the ideas underlying CDGP can also be applied using only user-provided training data, without formal specifications. We demonstrate the application of this method, called "informal CDGP," to software synthesis problems. Our results show that informal CDGP finds solutions faster (i.e. with fewer program executions) than standard GP. Additionally, we propose two new variants to informal CDGP, and find that one produces significantly more successful runs on about half of the tested problems. Finally, we study whether the addition of counterexample training cases to the training set is useful by comparing informal CDGP to using a static subsample of the training set, and find that the addition of counterexamples significantly improves performance.

Keywords: genetic programming · program synthesis · counterexamples · training data

1 Introduction

The bulk of the computational effort required for genetic programming (GP) is expended in the evaluation of programs in the evolving population. Typically, each program is evaluated on many inputs, which are generally referred to as "fitness cases" or "training cases". In most prior work, all available cases are used to evaluate each program.

Two recent developments in GP have offered new approaches to handling training cases that appear to provide significant advantages. One of these methods uses only a small, random sub-sample of the available cases each generation.

M. Giacobini et al. (Eds.): EuroGP 2024, LNCS 14631, pp. 21–37, 2024.
https://doi.org/10.1007/978-3-031-56957-9_2

This "down-sampling" saves significant computational effort per program evaluation, allowing one to run the evolutionary system for more generations with the same computational budget, leading to significant improvements in problem-solving power [8,15,19].

A second method, counterexample-driven genetic programming (CDGP), generates training cases using formal specifications that must be provided for the problem to be solved [1–3,23,24,32]. In particular, it is able to generate training cases that are not correctly solved by the evolving programs, adding these cases to a growing training set. These "counterexamples" provide more focused guidance to the evolutionary process than do random test cases, and appear to direct evolution more specifically to master aspects of the target problem that are not properly handled by individuals in the current population. While CDGP has been applied to constrained problem domains where it is possible to check whether any given program satisfies the given formal specifications, it is impossible to check whether programs over a Turing-complete language satisfy given formal specifications [21,27]. Therefore, CDGP cannot be applied directly to general program synthesis problems, where GP evolves programs that may include looping or recursion and access to potentially unbounded storage.

In this paper, we describe a novel method that builds on ideas of down-sampling the training data and CDGP by extending the idea of counterexamples to not require formal specifications.[1] The approach that we describe, "informal CDGP" (iCDGP), evaluates individuals during evolution using only a small sub-sample of the user-provided training cases, like down-sampled GP, allowing more individuals to be assessed within the same computational budget. When training cases are added to the training set, they are not chosen randomly, but rather are chosen to be counterexamples for the best individuals in the current population. This allows iCDGP to direct evolution in much the same way as CDGP, but without requiring that the user provide formal specifications for solutions to the target problem.

We test iCDGP on a set of general program synthesis benchmark problems, which require evolving programs in a Turing-complete language [14]. Initial experiments found that many times GP was not able to find a program that passed all training cases, meaning no new cases were added. We develop two new variants of iCDGP, and find that one in particular outperforms standard GP; this variant ensures that new cases are added to the training set throughout evolution, whether or not a program is found that passes the current training set. The second variant limits the size of the training set, motivated by making better use of the computational budget; this variant does not show as much empirical promise.

2   Related Work

This work takes its motivation from and builds on counterexample-driven GP and down-sampled lexicase selection. We describe each of those techniques in detail, and then discuss other related work.

---

[1] This paper expands on a poster paper that we published in GECCO 2020 [18].

2.1   Counterexample-Driven GP

CDGP uses specifications provided as formal constraints in order to generate the training cases used to evaluate a population of evolving programs [3,23,24]. CDGP was extended to use both formal constraints and user-provided training data to solve symbolic regression problems [1,2]. Additionally, CDGP has been combined with synthesis through unification, which allows it to partially decompose parts of problems into subproblems to solve [32].

CDGP evaluates individuals in the population against both a set of automatically generated training cases and the provided formal constraints. The training set is the primary method of evaluating individuals for parent selection, and the formal constraints are used to generate new training cases when necessary. The CDGP algorithm proceeds as follows: The training set is empty at the start of evolution. Then, each generation, every individual is evaluated on each training case in the training set. If any individual passes all of the training cases[2], CDGP uses a Satisfiability Modulo Theories (SMT) solver to test the individual on the problem's formal constraints. If the program passes the formal constraints, evolution stops because it has found a solution. If the individual fails a constraint, the SMT solver returns a counterexample in the form of a new case that the program does not pass. CDGP adds this case to the training set for the next generation. This process continues until it either finds a solution or reaches a maximum number of generations.

In standard CDGP, a new counterexample case is added to the training set only when a program passes all current training cases. However, one extension of CDGP adds a fitness threshold $q \in [0,1]$ that specifies the proportion of the training cases that an individual must pass before running the SMT solver on it to produce a new counterexample case [3]. This allows new cases to be added earlier, giving more search gradient for evolution to follow without having to find a program that passes all cases in the current training set. A value of $q = 1.0$ is equivalent to the standard CDGP, since it adds a new case only when an individual passes all current cases. On the basis of experimentation, the developers of this technique recommend a value for q in the range $[0.75, 1]$.

We are unaware of any previous work employing counterexamples without the use of formal specifications, as we do here.

2.2   Down-Sampled Lexicase Selection

While down-sampling of training data has been occasionally used in GP, it has recently been studied in the program synthesis domain when using lexicase parent selection [8,15,16,19]. In this setting, the training set is down-sampled to include a random subset of the training cases each generation. Down-sampled lexicase selection reduces the cost to evaluate each individual, with the same motivation as iCDGP.

---

[2] Every individual in the first generation passes the training set, since it is initially empty.

Multiple studies have examined why and where down-sampled lexicase performs well [16,20,28]. In evolutionary robotics, down-sampled lexicase selection has been used to limit the costs of robotics simulations [25,26]. More recently, work has been done to make informed decisions when selecting the cases that appear in the subsampled training set [4–6].

2.3 Other Related Work

Guiding learning with counterexamples that modify a training set has been recently explored in machine learning [7,29]. Implementations of this technique require an error table to be constructed from the model's misclassified data points from training, which is then used as specifications for how to construct counterexamples to train the model.

Metamorphic testing has been applied to GP to extend the usefulness of each case in exposing undesirable behaviors in candidate solutions without needing to include more cases to train on [30]. To apply metamorphic testing to a problem however, a user must first identify a metamorphic relation a solution program's output must exhibit; these metamorphic relationships are related to but different from the formal specifications required by CDGP.

The core of iCDGP has been used to develop human-driven genetic programming for program synthesis, in which a user is responsible for providing the initial training set and for verifying whether or not generated cases are counterexamples to a potential solution. A prototype system has shown promise on some basic program synthesis problems [9].

3   Informal Counterexample-Driven GP

Informal counterexample-driven GP (iCDGP) borrows motivation from CDGP, but deviates in some significant and novel ways. Specifically, we aim to adapt the core concept of a small training set that grows with added counterexamples. Since we do not have formal specifications, we instead expect the problem to be defined by a full training set of input/output examples, typically numbering 100 or more, which we call T.

In iCDGP, we use an active training set, $T_A \subseteq T$, that GP uses to evaluate the individuals in the population. In all of our experiments, T_A initially contains 10 random training cases from T, although other sizes could be used. During evolution, if an individual is found that passes all of the cases in T_A, we test the individual on all of the cases in $T \setminus T_A$; if it also passes all of them, then it is a training set solution and GP terminates. Otherwise, we select a random case in T that the individual does not pass, add it to T_A, and continue evolution. Note that if multiple individuals in a generation pass all of the cases in T_A, each of them goes through this process, potentially adding multiple new cases to T_A for the next generation.

Given that we already have a set of training cases, why does iCDGP use a smaller, likely less-informative set of active training cases? As with other approaches based on the sub-sampling of training cases (such as down-sampled lexicase and cohort lexicase selections [8, 19]), a smaller active training set allows iCDGP to perform fewer program executions per generation, making each generation computationally cheaper than if using the full set of training cases. In our experiments, we compare methods based on the same maximum number of program executions, allowing iCDGP to run for more generations than standard GP while using the same total program executions. Additionally, the CDGP idea of adding a counterexample case to T_A that the best individual does not pass allows it to augment the training set in ways that specifically direct GP to solve difficult parts of the problem that have not yet been solved by the evolving population.

In our experiments, we test several variants of iCDGP to try to improve it. One such variant uses a fitness threshold q to determine when to add a new counterexample, a variant introduced in formal CDGP [3]. For iCDGP, this triggers the system to test the individual on all of T, and then add a case to T_A that the individual does not pass. It is possible that the individual passes all of the cases in T that are not already in T_A; in this case, T_A does not change.

We designed another variant of iCDGP to address an issue that we discovered when analyzing our results, as presented in Sect. 5. In particular, many times iCDGP cannot find a program that passes all (or even a sufficiently high percentage to exceed a fitness threshold) of the active training set without passing all training cases. It may still be beneficial to add new training cases to T_A to provide GP with more information to guide search. Thus we created a variant of iCDGP, called generation-based case additions, that adds a new training case to T_A every d generations after the last case was added (whether through this process or an individual passing all of T_A). In order to select a case that provides better information for the search, we evaluate the best individual in the population (i.e. the one that passes the most cases in T_A, with ties broken at random) on all cases in T, and choose a random case that it does not pass.

Finally, we test a variant that sets a maximum number of cases to add to T_A. This variant prevents T_A from getting too large, which may be undesirable since it reduces the number of generations that GP can evaluate before using up the program execution budget. When adding a case that would increase the size of T_A past the given limit, we first remove the case in T_A that is passed by the most individuals in the population. In this way, we can remove cases that provide less useful direction to search while adding cases not passed by the best individuals.

4 Experiment Design

In this study we focus on general program synthesis problems, which require the GP system to generate programs that have similar qualities to the types of programs we expect humans to write. For our experiments we use 12 problems with a range of difficulties selected from the PSB1 benchmark suite [14]. These problems use different data types as inputs and outputs, and many require iteration or recursion and conditional execution to solve.

Table 1. PushGP system parameters.

Parameter	Value
population size	1000
max generations for runs using full training set	300
parent selection	lexicase
genetic operator	UMAD
UMAD addition rate	0.09
initial size of T_A	10

Each program synthesis problem is defined by a training set T of input/output examples, as well as an unseen test set that is used to test for generalization of solutions to unseen data.[3] As discussed above, when a program is found that passes all of the cases in the active training set T_A, it is tested on all cases in T; if it passes those as well, GP terminates. We then automatically simplify the program using a process that shrinks program sizes without changing its behavior on T; this simplification has been shown to increase generalization on the benchmark problems used in this study [10]. The simplified program then undergoes generalization testing on the test set; if it passes all of the unseen test cases, we consider it a successful run. If a program does not pass the test set, or if a run terminates from reaching the execution limit, it is marked a failure. We use a chi-square test with a 0.05 significance level to test for significant differences in success rates.

Our experiments use PushGP, which evolves programs in the Push programming language [31]. Push, designed specifically for use in GP, uses a set of typed stacks to store data manipulated by a program. Push programs are hierarchical lists containing data literals, which are pushed onto stacks when encountered in programs, and instructions, which take their inputs from specifics stacks and return their results to the stacks. We use an implementation of PushGP written in Clojure in our experiments.[4]

The PushGP system parameters used in our experiments are given in Table 1. Individual genomes are stored in the Plushy representation, and translated into Push programs for execution. We use uniform mutation with additions and deletions (UMAD) as our only genetic operator, making all children through mutation only, since this mutation has produced the best known results when using PushGP on these benchmark problems [13]. We use the size-neutral version of UMAD, adding new instructions before or after 9% of instructions in the parent. Additionally, we use lexicase selection for parent selection [14, 17].

Lexicase selection has one peculiar characteristic with respect to iCDGP: when an individual is found that passes all cases in T_A but not all those in T, we

---

[3] The datasets for these problems are available at https://github.com/thelmuth/program-synthesis-benchmark-datasets.

[4] https://github.com/lspector/Clojush.

Table 2. Full training set size and program execution limit for each problem.

Problems	Training Size	Executions
Compare String Lengths, Double Letters, Mirror Image, Replace Space With Newline, Smallest, String Lengths Backwards, Syllables	100	30,000,000
Last Index of Zero, X-Word Lines	150	45,000,000
Negative to Zero, Scrabble Score	200	60,000,000
Vector Average	250	75,000,000

add a new case to T_A. However, the selection of parents for the next generation is based on T_A before the new case is added, since that is the set of cases that the population is evaluated on. In lexicase selection, if an individual passes all cases considered for selection and no other individual does, then it will be selected in every single parent selection event that generation. Thus we might expect substantial drops in population diversity each time we add a new case to T_A. We investigate the implications of this interaction between lexicase selection and iCDGP empirically in Sect. 5.4.

Since iCDGP executes fewer programs per generation, all of our PushGP runs are limited by the number of program executions they allow, equivalent to using a population size of 1000 and 300 maximum generations for a full training set. This ensures that all methods receive the same number amount of computation. Since iCDGP uses fewer training cases to evaluate each individual, it runs for more generations to make up the same number of program executions. The number of training cases in the full training set varies per problem, so the maximum program execution limits also vary per problem, and both are given in Table 2.

5   Results

We first present results comparing iCDGP to GP using the full training set. The first three columns of Table 3 give the number of successful GP runs out of 100 using a full training set, iCDGP, and iCDGP with a fitness threshold of $q = 0.8$. First, note that iCDGP performed a bit worse than using the full training set, including significantly worse on four problems while only significantly better on one. On the other hand, iCDGP using a fitness threshold performed significantly better than the full training set on three problems while only performing significantly worse on two, showing the benefits of adding cases before finding a solution on the training set.

Despite producing no notable improvement in performance on these benchmark problems over using the full training set, we did notice that the solutions that iCDGP found often occurred earlier in evolutionary time than with the full training set. For example, Fig. 1 shows the cumulative number of successes over evolutionary time on the Vector Average problem; this plot is representative of many of the other problems we observed. Both of the iCDGP methods (with

Table 3. The number of successes out of 100 GP runs. All results are compared to **Full**; results that are significantly better are in **bold**, and results that are significantly worse are *underlined and italicized.* **Full** is GP using the full training set. **iCDGP** is the standard version of iCDGP. **q = 0.8** is the variant of iCDGP that adds a case any time an individual passes more cases than the fitness threshold of $q = 0.8$. The three columns labeled with values for **d** are the variant that adds a case to the training set every d generations.

Problem	Full	iCDGP	q = 0.8	d = 25	d = 50	d = 100
CSL	32	*13*	41	30	20	*18*
DL	19	24	26	**37**	32	29
LIoZ	62	*41*	56	63	62	65
MI	100	98	*89*	*93*	98	96
NTZ	80	76	80	79	80	81
RSWN	87	94	**96**	91	**96**	91
SS	13	15	**30**	**50**	**31**	**42**
Smallest	100	*93*	*93*	96	95	95
SLB	94	90	90	*83*	87	87
Syl	38	*24*	44	**69**	**62**	49
VA	88	87	89	**97**	**97**	**98**
XWL	61	**75**	**82**	**85**	**89**	**87**

and without a fitness threshold) find many solutions quite early in their runs, reaching 50 successes around 10 million program executions, at which point the full training set has produced only 14 solutions. However, the full training set catches up over evolutionary time, reaching about the same number of successes by the time it hits the maximum number of program executions. iCDGP's ability to find solutions earlier likely stems from it executing many fewer programs per generation, allowing it to produce more generations (and therefore explore more programs) within the same number of program executions. The rapid production of solutions provides one argument for using iCDGP.

5.1   Variant: Generation-Based Case Additions

With generation-based case additions, we add a new case to T_A every time d generations have passed without a new case being added otherwise. We tested three settings for d: 25, 50, and 100 generations. Note that failed iCDGP runs often finished after 1000 to 3000 generations, depending on how many cases are added to T_A.

The last three columns of Table 3 present the number of successful runs for different settings of d. iCDGP with generation-based additions performed similarly to or better than iCDGP without them on every problem for all three settings of d. While all three performed sometimes better and sometimes worse than each other, we will concentrate on $d = 50$ here, which was significantly better than the full training set on five problems while never performing significantly worse.

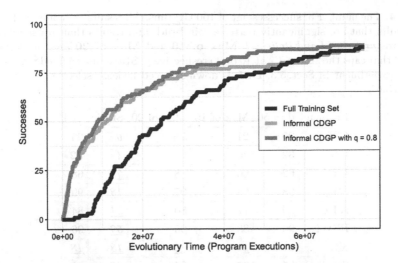

Fig. 1. Cumulative number of successful GP runs on the Vector Average problem over evolutionary time, as measured by program executions.

Generation-based case additions turns iCDGP's questionable benefits into clear ones. They also present an improvement over using a fitness threshold, likely because they allow for the addition of new cases without an individual having to reach the threshold. Future work would be needed to determine whether selecting a new case that is not passed by the best individual helps, or if adding any random case from $T \setminus T_A$ would be sufficient.

5.2   Variant: Maximum Size of Active Training Set

Since it appears that some of the benefits of iCDGP derive from the fact that it uses a small number of cases per program evaluation, we to hypothesize that its performance might improve if the number of cases were capped. For the experiments that produced the results shown in the columns **Max of 10** and **Max of 20** in Table 4, we began with the version of iCDGP using generation-based case additions every 50 generations. To this configuration we added a mechanism that removes a case each time a new case is added, once the number of cases has reached a pre-specified maximum. Specifically, whenever we add a case that would increase the size of T_A past the given limit, we first remove the case in T_A that is passed by the most individuals in the current population, with the intention to remove cases that provide less useful direction to search.

As can be seen in Table 4, limiting T_A to 10 cases degrades problem-solving significantly on the Scrabble Score and Syllables problems. Limiting T_A to 20 cases produces significantly worse results on Scrabble Score, but significantly better on Compare String Lengths. Neither of these results suggests that limiting the size of T_A deserves recommendation.

Table 4. The number of successes out of 100 GP runs. All results are compared to $\mathbf{d} =$ **50**; results that are significantly better are in **bold**, and results that are significantly worse are _underlined and italicized._ **Max of 10** and **Max of 20** are the variant of iCDGP that caps the size of T_A at 10 or 20 respectively. **Static** uses a fixed training set, for the experiment in Sect. 5.3. **DSL** is down-sampled lexicase selection, as discussed in Sect. 5.6.

Problem	d = 50	Max of 10	Max of 20	Static	DSL
CSL	20	21	**53**	_0_	25
DL	32	46	37	_4_	**72**
LIoZ	62	66	58	_7_	68
MI	98	98	97	_13_	99
NTZ	80	77	80	_31_	84
RSWN	96	88	95	_57_	96
SS	31	_2_	_15_	_13_	_18_
Smallest	95	97	94	_40_	99
SLB	87	93	85	_35_	**96**
Syl	62	_36_	52	_9_	61
VA	97	95	95	_71_	100
XWL	89	91	94	_35_	95

5.3   Benefits of Counterexample Cases

One may wonder whether the benefits we have demonstrated with iCDGP come entirely from having a small active training set T_A on which we evaluate each individual, reducing the number of program executions per generation. In other words, it is possible that adding counterexample cases to T_A provides no benefits. To test this hypothesis, we conducted a set of runs that use a static active training set consisting of 10 random cases, the same number as the size of T_A at the start of our iCDGP runs. Note that the only functional difference in these methods happens when a program is found that passes all cases in T_A. When a run with a static training set finds a program that passes all cases in T_A, it is simply tested for generalization.

Table 4 compares the number of successes produced by GP with iCDGP adding a case every $d = 50$ generations to GP with a static training set. iCDGP is significantly better on every problem, often by huge margins. These differences highlight the importance of iCDGP's additions of counterexample cases to T_A.

5.4   Effects on Population Diversity

We are interested in the effects of iCDGP, especially with lexicase selection, on population diversity. In particular, as we discussed in Sect. 4, when an individual passes all cases in T_A and iCDGP adds another case, lexicase selects that individual as the parent of every child in the next generation. If more than one such

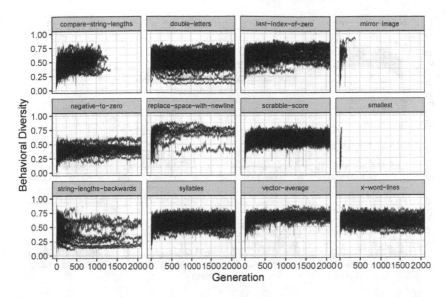

Fig. 2. Population behavioral diversity for standard iCDGP runs, cropped at 2000 generations. Each run is plotted separately. Note that all runs for Mirror Image and Smallest found solutions early in the runs, and Compare String Lengths runs ended earlier than others because they often added many training cases to T_A.

individual is found in the same generation, then the selections will be randomly distributed among them.

These *hyperselection* events will mean that every individual created after such an event will be a descendant of the hyperselected individual. Previous work studied hyperselection events with lexicase selection, but specifically individuals that received 5–10% of the selections in a generation, not 100% of them [12]. This work found that hyperselection events had no noticeable effects on problem-solving performance, and only caused brief reductions in population diversity. Another study of lexicase selection found that it is able to quickly recover population diversity in situations when the population had low diversity [11]. Here we examine whether these hyperselection events have detrimental effects on population diversity when using iCDGP.

We measure population diversity in terms of *behavioral diversity*, or the proportion of distinct behavior vectors produced by a population [22]; a *behavior vector* is a list of outputs that the program produces when run on the cases in T_A. Figure 2 plots the behavioral diversity of every single iCDGP run on all 12 problems as a separate line. Looking closely, there are clearly instances where population diversity drops drastically in one generation[5], with many in the Scrabble Score and Vector Average problems, and a few in most of the other

---

[5] Note that the diversity does not go all the way to 0, since even if one parent created all of the children in the next generation, some of those children are likely to display different behaviors from the parent.

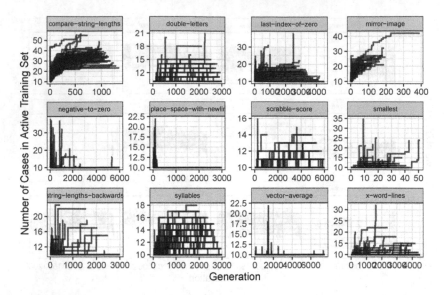

Fig. 3. The number of training cases in the active training set T_A for iCDGP runs. Each run is plotted separately. Note that no cases are ever removed, so each line can only increase. Also note different x-axis and y-axis scales per problem.

problems. Most of these drops in diversity follow one of two patterns: a solution to the full training set T is found in the next generation, leading to a line that drops down and then ends; or a quick increase in diversity over a few generations back to levels seen before the drop. On the other hand, we see little evidence for sudden drops in diversity leading to extended stretches of low diversity.

So, while these hyperselection phenomena do occur when an individual passes all cases in T_A, there does not seem to be corresponding long-term detrimental effects on population diversity. Lexicase selection may be the cause and the cure, as its case-by-case effects provide boosts in diversity following hyperselection.

5.5   Number of Active Cases

In order to get a better idea of how often iCDGP adds cases to T_A, we plot the number of cases in T_A over evolutionary time for standard iCDGP in Fig. 3 and for the version that adds a case every 50 generations in Fig. 4.

For iCDGP, we see many different patterns of when and how many cases are added to T_A. For example, Compare String Lengths and Mirror Image are the only problems in our benchmark set with Boolean-valued outputs, making them easier to pass all cases in T_A without passing all of T than problems with outputs coming from a wider domain, such as numbers or strings. Thus we find it unsurprising that these two problems consistently see the largest growth in T_A. Other problems add cases at different rates, with Replace Space with Newline, Vector Average, and Negative to Zero falling at the other extreme, where a few

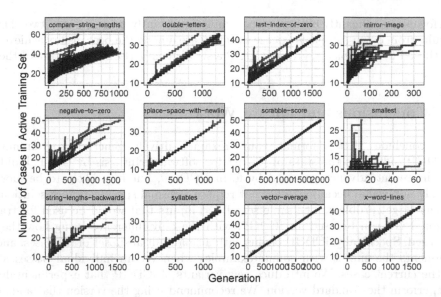

Fig. 4. The number of training cases in the active training set T_A for iCDGP with $d = 50$, adding a new case every 50 generations. Each run is plotted separately. Note that no cases are ever removed, so each line can only increase. Also note different x-axis and y-axis scales per problem.

runs added quite a few cases early and were solved, while the rest never added any cases.

The stair-step pattern of sizes of T_A in Fig. 4 reflects the cases that are added to T_A after 50 generations since a case was lasted added. Some problems, corresponding roughly with the problems in Fig. 3 that add few cases, rarely if ever add a case besides every 50 generations. Other problems still seem to add quite a few cases for individuals that pass all of T_A. We find no correlation between these two types of problems and those at which this version of iCDGP performs better compared to the standard. The performance improvement seen when adding a case every 50 generations seems to benefit both kinds of problems.

5.6   Comparison with Down-Sampled Lexicase Selection

We compare iCDGP to down-sampled lexicase selection, using results from [15]. To ensure fairness of the comparison, we only consider down-sampled lexicase with down-sample rates which result in 10 cases being evaluated each generation, which is the size of iCDGP's active set T_A. For instances where no such down-sampling rate existed (Last Index of Zero, Vector Average, and X-Word Lines), we used the results from a down-sampling rate that resulted in just a few more than 10 cases being evaluated per generation.

We found that iCDGP (using $d = 50$) is competitive to down-sampled lexicase selection, with a comparison in Table 4. Of the 12 test problems, iCDGP performed significantly better on one, while significantly worse on only two. The results from down-sampled lexicase selection are among the best results achieved on these PSB1 problems, giving iCDGP a strong comparison to the state-of-the-art.

6   Conclusions and Future Work

We conclude that informal counterexample-driven genetic programming (iCDGP) advances the state of the art for software synthesis by GP. It builds on the recent advance provided by formal CDGP, but it is likely to be more widely applicable because it does not require a formal specification of solutions to the target problem. The same set of test inputs that would be used for traditional GP can be used for iCDGP, with the only difference being how they are used. Specifically, iCDGP begins with a small initial subset of the cases, and augments the subset with counterexamples whenever an individual passes all of the current cases. We introduce new variants of iCDGP that experimentally outperform the standard version. We recommend using the version that adds a new case to the active set T_A every d generations, ensuring that cases are added even if no program is found that passes all cases in T_A. This variant performed best for iCDGP, and future work could investigate its use in CDGP with formal constraints.

Although we explored several variants of iCDGP, we anticipate other variants to emerge which may outperform ones presented here. Future work should focus on conducting further analyses of the underlying evolutionary dynamics that are responsible for the success of the technique to guide us in developing improvements. We have presented here some preliminary data on behavioral diversity and numbers of cases in T_A over evolutionary time, but many other aspects of these runs can be investigated, and other variants of the technique tested to explore hypotheses about the reasons that it works. For example, with respect to generation-based additions, it would be useful to learn wither it is important to include new cases that are not passed by the best individual, or if the same benefit would result, more simply, from adding any random case from $T \setminus T_A$.

Acknowledgements. We thank the members of the PUSH lab for discussions that improved this work. This material is based upon work supported by the National Science Foundation under Grant No. 2117377. Any opinions, findings, and conclusions or recommendations expressed in this publication are those of the authors and do not necessarily reflect the views of the National Science Foundation.

References

1. Błądek, I., Krawiec, K.: Solving symbolic regression problems with formal constraints. In: GECCO 2019: Proceedings of the Genetic and Evolutionary Computation Conference, pp. 977–984. ACM, Prague (2019). https://doi.org/10.1145/3321707.3321743

2. Błądek, I., Krawiec, K.: Counterexample-driven genetic programming for symbolic regression with formal constraints. IEEE Trans. Evol. Comput. **27**(5), 1327–1339 (2023). https://doi.org/10.1109/TEVC.2022.3205286

3. Błądek, I., Krawiec, K., Swan, J.: Counterexample-driven genetic programming: heuristic program synthesis from formal specifications. Evol. Comput. **26**(3), 441–469 (2018). https://doi.org/10.1162/evco_a_00228

4. Boldi, R., et al.: The problem solving benefits of down-sampling vary by selection scheme. In: Proceedings of the Companion Conference on Genetic and Evolutionary Computation, pp. 527–530 (2023)

5. Boldi, R., et al.: Informed down-sampled lexicase selection: identifying productive training cases for efficient problem solving (2023). https://doi.org/10.48550/arXiv.2301.01488, arXiv:2301.01488

6. Boldi, R., Lalejini, A., Helmuth, T., Spector, L.: A static analysis of informed down-samples. In: Proceedings of the Companion Conference on Genetic and Evolutionary Computation, GECCO 2023 Companion, pp. 531–534. Association for Computing Machinery, New York (2023). https://doi.org/10.1145/3583133.3590751

7. Dreossi, T., Ghosh, S., Yue, X., Keutzer, K., Sangiovanni-Vincentelli, A., Seshia, S.A.: Counterexample-guided data augmentation (2018)

8. Ferguson, A.J., Hernandez, J.G., Junghans, D., Lalejini, A., Dolson, E., Ofria, C.: Characterizing the effects of random subsampling on lexicase selection. In: Banzhaf, W., Goodman, E., Sheneman, L., Trujillo, L., Worzel, B. (eds.) Genetic Programming Theory and Practice XVII. GEC, pp. 1–23. Springer, Cham (2020). https://doi.org/10.1007/978-3-030-39958-0_1

9. Helmuth, T., Frazier, J.G., Shi, Y., Abdelrehim, A.F.: Human-driven genetic programming for program synthesis: a prototype. In: Proceedings of the Companion Conference on Genetic and Evolutionary Computation, GECCO 2023 Companion, pp. 1981–1989. Association for Computing Machinery, New York (2023). https://doi.org/10.1145/3583133.3596373

10. Helmuth, T., McPhee, N.F., Pantridge, E., Spector, L.: Improving generalization of evolved programs through automatic simplification. In: Proceedings of the Genetic and Evolutionary Computation Conference, GECCO 2017, pp. 937–944. ACM, Berlin (2017). https://doi.org/10.1145/3071178.3071330, http://doi.acm.org/10.1145/3071178.3071330

11. Helmuth, T., McPhee, N.F., Spector, L.: Effects of lexicase and tournament selection on diversity recovery and maintenance. In: GECCO 2016 Companion: Proceedings of the Companion Publication of the 2016 Annual Conference on Genetic and Evolutionary Computation, pp. 983–990. ACM, Denver (2016). https://doi.org/10.1145/2908961.2931657, http://doi.acm.org/10.1145/2908961.2931657

12. Helmuth, T., McPhee, N.F., Spector, L.: The impact of hyperselection on lexicase selection. In: Friedrich, T. (ed.) GECCO 2016: Proceedings of the 2016 Annual Conference on Genetic and Evolutionary Computation, pp. 717–724. ACM, Denver (2016). https://doi.org/10.1145/2908812.2908851

13. Helmuth, T., McPhee, N.F., Spector, L.: Program synthesis using uniform mutation by addition and deletion. In: Proceedings of the Genetic and Evolutionary Computation Conference, GECCO 2018, pp. 1127–1134. ACM, Kyoto (2018). https://doi.org/10.1145/3205455.3205603, http://doi.acm.org/10.1145/3205455.3205603

14. Helmuth, T., Spector, L.: General program synthesis benchmark suite. In: GECCO 2015: Proceedings of the 2015 Annual Conference on Genetic and Evolutionary Computation, pp. 1039–1046. ACM, Madrid (2015). https://doi.org/10.1145/2739480.2754769, http://doi.acm.org/10.1145/2739480.2754769

15. Helmuth, T., Spector, L.: Explaining and exploiting the advantages of down-sampled lexicase selection. In: Artificial Life Conference Proceedings, pp. 341–349. MIT Press (2020). https://doi.org/10.1162/isal_a_00334, https://www.mitpressjournals.org/doi/abs/10.1162/isal_a_00334

16. Helmuth, T., Spector, L.: Problem-solving benefits of down-sampled lexicase selection. Artif. Life **27**(3–4), 183–203 (2022). https://doi.org/10.1162/artl_a_00341

17. Helmuth, T., Spector, L., Matheson, J.: Solving uncompromising problems with lexicase selection. IEEE Trans. Evol. Comput. **19**(5), 630–643 (2015). https://doi.org/10.1109/TEVC.2014.2362729, http://ieeexplore.ieee.org/stamp/stamp.jsp?tp=&arnumber=6920034

18. Helmuth, T., Spector, L., Pantridge, E.: Counterexample-driven genetic programming without formal specifications. In: Proceedings of the 2020 Genetic and Evolutionary Computation Conference Companion, GECCO 2020, pp. 239–240. ACM (2020). https://doi.org/10.1145/3377929.3389983

19. Hernandez, J.G., Lalejini, A., Dolson, E., Ofria, C.: Random subsampling improves performance in lexicase selection. In: GECCO 2019: Proceedings of the Genetic and Evolutionary Computation Conference Companion, pp. 2028–2031. ACM, Prague (2019). https://doi.org/10.1145/3319619.3326900

20. Hernandez, J.G., Lalejini, A., Ofria, C.: An exploration of exploration: measuring the ability of lexicase selection to find obscure pathways to optimality. In: Banzhaf, W., Trujillo, L., Winkler, S., Worzel, B. (eds.) Genetic Programming Theory and Practice XVIII. Genetic and Evolutionary Computation, pp. 83–107. Springer, Singapore (2022). https://doi.org/10.1007/978-981-16-8113-4_5

21. Hopcroft, J.E., Ullman, J.D.: Introduction to Automata Theory, Languages, and Computation. Addison-Wesley (1979)

22. Jackson, D.: Promoting phenotypic diversity in genetic programming. In: Schaefer, R., Cotta, C., Kołodziej, J., Rudolph, G. (eds.) PPSN 2010. LNCS, vol. 6239, pp. 472–481. Springer, Heidelberg (2010). https://doi.org/10.1007/978-3-642-15871-1_48

23. Krawiec, K., Błądek, I., Swan, J.: Counterexample-driven genetic programming. In: Proceedings of the Genetic and Evolutionary Computation Conference, GECCO 2017, pp. 953–960. ACM, Berlin (2017). https://doi.org/10.1145/3071178.3071224, http://doi.acm.org/10.1145/3071178.3071224

24. Krawiec, K., Błądek, I., Swan, J., Drake, J.H.: Counterexample-driven genetic programming: stochastic synthesis of provably correct programs. In: Lang, J. (ed.) Proceedings of the Twenty-Seventh International Joint Conference on Artificial Intelligence (IJCAI-18), pp. 5304–5308. International Joint Conferences on Artificial Intelligence, Stockholm (2018). https://doi.org/10.24963/ijcai.2018/742, https://www.ijcai.org/proceedings/2018/742

25. Moore, J.M., Stanton, A.: Lexicase selection outperforms previous strategies for incremental evolution of virtual creature controllers. In: Proceedings of the European Conference on Artificial Life, pp. 290–297 (2017). https://doi.org/10.1162/ecal_a_0050_14, https://www.mitpressjournals.org/doi/abs/10.1162/ecal_a_0050_14

26. Moore, J.M., Stanton, A.: Tiebreaks and diversity: isolating effects in lexicase selection. In: The 2018 Conference on Artificial Life, pp. 590–597 (2018). https://doi.org/10.1162/isal_a_00109, https://www.mitpressjournals.org/doi/abs/10.1162/isal_a_00109

27. Rice, H.G.: Classes of recursively enumerable sets and their decision problems. Trans. Am. Math. Soc. **74**(2), 358–366 (1953). http://www.jstor.org/stable/1990888

28. Schweim, D., Sobania, D., Rothlauf, F.: Effects of the training set size: a comparison of standard and down-sampled lexicase selection in program synthesis. In: 2022 IEEE Congress on Evolutionary Computation (CEC), pp. 1–8 (2022). https://doi.org/10.1109/CEC55065.2022.9870337
29. Sivaraman, A., Farnadi, G., Millstein, T., den Broeck, G.V.: Counterexample-guided learning of monotonic neural networks (2020)
30. Sobania, D., Briesch, M., Röchner, P., Rothlauf, F.: MTGP: combining metamorphic testing and genetic programming (2023)
31. Spector, L., Klein, J., Keijzer, M.: The push3 execution stack and the evolution of control. In: GECCO 2005: Proceedings of the 2005 conference on Genetic and evolutionary computation, vol. 2, pp. 1689–1696. ACM Press, Washington DC (2005). https://doi.org/10.1145/1068009.1068292, http://www.cs.bham.ac.uk/wbl/biblio/gecco2005/docs/p1689.pdf
32. Welsch, T., Kurlin, V.: Synthesis through unification genetic programming. In: Proceedings of the 2020 Genetic and Evolutionary Computation Conference, GECCO 2020, pp. 1029–1036. Association for Computing Machinery, New York (2020). https://doi.org/10.1145/3377930.3390208, https://doi.org/10.1145/3377930.3390208

Investigating Premature Convergence in Co-optimization of Morphology and Control in Evolved Virtual Soft Robots

Alican Mertan[✉][iD] and Nick Cheney[iD]

University of Vermont, Burlington, VT 05401, USA
{alican.mertan,ncheney}@uvm.edu

Abstract. Evolving virtual creatures is a field with a rich history and recently it has been getting more attention, especially in the soft robotics domain. The compliance of soft materials endows soft robots with complex behavior, but it also makes their design process unintuitive and in need of automated design. Despite the great interest, evolved virtual soft robots lack the complexity, and co-optimization of morphology and control remains a challenging problem. Prior work identifies and investigates a major issue with the co-optimization process – fragile co-adaptation of brain and body resulting in premature convergence of morphology. In this work, we expand the investigation of this phenomenon by comparing learnable controllers with proprioceptive observations and fixed controllers without any observations, whereas in the latter case, we only have the optimization of the morphology. Our experiments in two morphology spaces and two environments that vary in complexity show, concrete examples of the existence of high-performing regions in the morphology space that are not able to be discovered during the co-optimization of the morphology and control, yet exist and are easily findable when optimizing morphologies alone. Thus this work clearly demonstrates and characterizes the challenges of optimizing morphology during co-optimization. Based on these results, we propose a new body-centric framework to think about the co-optimization problem which helps us understand the issue from a search perspective. We hope the insights we share with this work attract more attention to the problem and help us to enable efficient brain-body co-optimization.

Keywords: Evolutionary robotics · Soft robotics · Brain-body co-optimization

1 Introduction

Evolving virtual creatures is a field with a long history, starting with Karl Sims' seminal work "Evolving Virtual Creatures" [51] almost 30 years ago. Karl Sims' approach of co-optimizing brain and body then widely adapted and researched, especially in soft robotics [2–6, 10, 20, 38, 40, 54].

M. Giacobini et al. (Eds.): EuroGP 2024, LNCS 14631, pp. 38–55, 2024.
https://doi.org/10.1007/978-3-031-56957-9_3

The field of soft robotics with volumetric actuation started with [13,15,55] and accelerated with the availability of simulators [2,14,33,37]. The compliance and flexibility of the material make soft robots capable of exhibiting complex and unintuitive behaviors, resulting in different abilities of soft robots such as walking [2,5,6,10,20,22,24–26,38–40,43,52], swimming [9,20], squeezing through obstacles [4], damage recovery and regeneration [16,17,28], shape change [48] and self-replication [23] being tested. The highly non-linear and complex nature of soft material dynamics that don't exist in their rigid counterparts [21,46,49], provides an increased potential for morphological computation [41,42] that can be unlocked by a tightly integrated body design and control strategy.

The ability to output complex behavior, however, makes the design process of soft robots counter-intuitive, highly encouraging automated design over manual design. Despite the great interest in the automated brain-body co-optimization [2–6,9,10,13,15,20,22,26,35,38,40,53,54], evolved soft robots struggles to surpass the complexity of the Sims' initial creatures [51]. Prior work identifies and investigates an important phenomenon that hinders the co-optimization process – premature convergence of morphology [3,20].

As the optimization takes place, two parts of the solution, the brain and the body, become more and more specialized for each other, making the overall performance of the solution very sensitive to changes in either component [3]. In the case of co-optimizing the brain and body in soft robots, this is especially prominent for the morphology of the solution where even small changes in the morphology can drastically reduce the performance of the solution [40]. Commonly used algorithms for brain-body co-optimization don't like solutions with poor performances and focus the search over other parts of the morphology space, making them unable to discover high-performing regions in the search space. To overcome the poor search over morphologies, prior work proposes the use of better search algorithms that reduce the selection pressure of individuals with new body plans to promote more search over morphologies [5,30], or proposes ways to alleviate the performance decline during the morphological search by using different genetic representations for coordinated changes [54,57] or by using controllers that are robust to changes in the morphology [40].

Rather than proposing a solution, we provide an investigation of this phenomenon following the previous investigation of Cheney et al. [3]. Specifically, we investigate the use of neural network controllers with proprioceptive observations where we optimize both morphology and control, and heuristically devised fixed controllers where we only optimize the morphology. This allows us to shed light on the effects of control optimization and specialization on the co-optimization process and understand the premature convergence issue. The main contributions of this work are to:

- expand previous investigations towards a more generalized understanding of the brain-body fragile co-optimization phenomenon by focusing on important and missing aspects such as the use of more varied and complex controllers, sensory information, morphology spaces, and environments (Sect. 3).

– provide arguably the most concrete example that shows the existence of high-performing designs that aren't discovered by the co-optimization process, due to poor search over the morphology space (Sect. 4).
– develop a new body-centric framework to conceptualize the fitness landscape of the co-optimization problem, allowing us to understand the challenge of co-optimization from a search perspective (Sec. 5).
– organize the existing solutions to premature convergence within the proposed framework, helping us to clarify promising research directions (Sect. 5).

We hope these findings and the proposed framework will be useful for further research into enabling better brain-body co-optimization.

2   Experiment Design

2.1   Simulation

We perform our experiments in the Evolution Gym version 1.0.0 (Evogym) [2]. It is an open-source voxel-based soft-body simulator with a suite of benchmark tasks. Voxels are represented as a mass-spring system, and it works in 2D, similar to the simulation engines in [11, 20, 36–40, 43, 54].

Each voxel of the robot can have a different type that determines its behavior and is represented with different colors visually. There are two types of active materials that can actuate volumetrically, either horizontally or vertically, and they are under the direct control of the controller. These materials are controlled by specifying the target length $a \in [0.6, 1.6]$ (in the form of a multiplier of the resting length) at each time step where materials gradually expand/contract to achieve the target length. We query the controller every 5th timestep (referred to as the effective timestep) and repeat the queried action until the next effective time step to limit high-frequency dynamics that might lead to unstable behavior. Additionally, there are two types of passive materials that are not under direct control, and they differ in their elasticity, one being **rigid** and the other being elastic. Designing a robot is simply choosing the existence and material type of voxels in a grid layout. While Evogym allows for specifying whether neighboring voxels are connected to each other, we omit this feature and assume that neighboring voxels are always connected to each other.

The simulation engine provides various proprioceptive and environmental observations. We only use proprioceptive observations of volume, speed, and material properties of voxels and a global time signal.

2.2   Task and Environments

We use two environments with a locomotion task that are presented in the Evogym benchmark suite, namely Walker-v0 and BridgeWalker-v0 [2]. Locomotion is the most used task for voxel-based soft robotics research [3, 5–7, 10, 11, 16, 17, 24–28, 35, 36, 38–40, 43, 52–54]. We adopt the modified fitness function in [40], $R(r, T) = \Delta p_x^r + \mathbf{I}(r) + \sum_{t=0}^{T} -0.01 + 5$, which rewards robots r for moving in the positive x direction with the term Δp_x^r, rewards them for finishing the task with the indicator function $\mathbf{I}(p_x^r)$, and encourages them to finish the task faster by applying a negative penalty at each time step with the term $\sum_{t=0}^{T} -0.01$. The last term, $+5$, is equal in magnitude to the maximum penalty and is used to shift the reward to be positive for ease of analysis.

Environments often consist of a flat [3,5–7,10,16,17,24–28,35,36,38–40,52–54] or uneven [6,11,43,52] surfaces, supposedly harder task. We experiment with both locomotion on a flat surface (Walker-v0) and an uneven, dynamic surface (BridgeWalker-v0) to understand the relationship between the environment and the brain-body co-optimization process.

2.3   Robot Design and Controller Strategies

Robot Representation and Design. Following [2,38,40], we use a direct representation for robot design where the existence of voxels and their material type encoded in a matrix $\in M_{h \times w}(T)$, where $T \in \{0,1,2,3,4\}$. While indirect encodings are commonly used in prior work that works in 3D [3–6,9,10,20,22,25], working in 2D makes it possible to use direct encoding which is shown to be comparable to complex indirect encodings [2] and allows us to control for the change in morphology during mutation. Following the common practice of limiting the morphology space [3,5,10,24,35,38,40,54], we experiment with two morphology spaces, $(h, w) \in \{(5,5), (7,7)\}$, to provide more instances for investigation of premature convergence, as encouraged by previous investigation [3].

Controller Design and Model. We compare a learnable controller and a fixed, non-sensing controller. The fixed controller allows us to create a scenario where fragile co-adaptation of design and controller cannot occur due to the lack of optimization on the controller part and helps us investigate premature convergence that we observe with learnable controllers. Specifically, we use a modular/decentralized control strategy as a learnable control[1] as they are shown to help with the brain-body co-optimization [40].

Learnable Controller. One of the common choices in control design is the use of modular control strategy [18,36,38,40,43]. They are considered to be more versatile and robust to changes in the robot morphology [59], which helps with the co-optimization process [40], and are compatible with any robot morphology.

The modular controller consists of a shared neural network model (single hidden layer with 32 neurons) assigned to each active voxel, observing a local patch around it and only determining the behavior of a single voxel it is assigned. The controller observes a 3×3 window centered around the active voxel. The observations to the controller consist of the proprioceptive information (volume, speed, and material type) of the voxels in this observation window and timestep mod 2 as the time signal to have a fair comparison with the fixed controller.

---

[1] We repeated every experiment with the most commonly used global/centralized control strategy [11,18,36,39,40,52,53] and didn't observe any qualitatively different results. For the sake of space and simplicity, we omit the global controller and only present the results with the modular controller.

Fixed Controller. One of the simplest control strategies used in the literature is the fixed actuation following a global signal [3–6,9,22,24–28,53]. While the learnable controller performs a nonlinear mapping from proprioceptive observations to actions, fixed controller lacks any sensory input. Active voxels under the control of the fixed controller alternate between expanding and contracting maximally at every effective timestep. Therefore, the controller has zero learnable parameters and doesn't require any control optimization, hence the name fixed. Having a fixed controller that doesn't require any control optimization allows us to create a scenario where fragile co-adaptation doesn't occur, helping us understand premature convergence with learnable controllers better.

2.4   Optimization Algorithm

To optimize the design and control of the soft robots, we adopt the use of evolutionary algorithms, similar to [3,5,20,38,40,44,54,56,57]. Evolutionary algorithms allow us to simultaneously make improvements in both design and control, supposedly reducing the computational cost of two-level optimization approaches where outer loop does design optimization and the inner loop optimizes the control for a given design, as in [2].

In particular, we use age-fitness Pareto optimization (AFPO) [47] with truncation selection. Individuals' ages are increased at every generation, and we inject a random individual at each generation with the age of 0 for diversity. Recombination wasn't considered and the offspring are created through mutation only. Individuals created by mutation inherit the age of their parent.

Following [2,40], we use a mutation operator that creates new designs by changing each voxel type of a robot with a 10% probability from/to an empty voxel. To mutate in the control space, we add noise sampled from $\mathcal{N}(0, 0.1)$ to all learnable controller parameters.

When creating offspring through mutation, we either mutate the body plan or the controller of the solution. Following [5,40], there is a 50% probability of choosing either component of the solution for the mutation. Note that the fixed controller doesn't have any controller parameters. Yet, for a fair comparison, individuals with fixed controllers go through control mutation with the same probability, effectively creating a copy of the same solution.

3   Co-optimization

The main problem we are interested in is the co-optimization of morphological design and control, and understanding the premature convergence of morphology. To compare the effects of different control strategies, we evolve two populations of solutions: one with learnable and one with fixed controllers. While with the learnable controller, we have the co-optimization of morphology and control, the fixed controller allows us to test the case where we only optimize the morphology, allowing us to investigate the effect of controllers on the premature convergence issue. We experiment with evolving these populations under 4

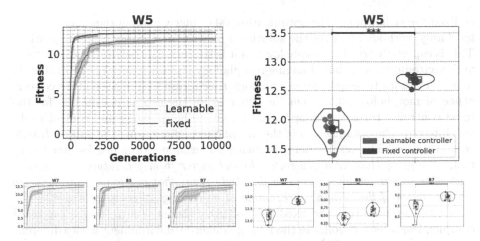

Fig. 1. Fitness over time plots (left) and distributions of the best solutions (right) for the co-optimization experiments. The simplest controller, the fixed controller, significantly outperforms the learnable controller in all experimental settings. Moreover, populations with fixed controllers converge faster. Left: Solid lines show the best fitness found at each generation, averaged across 10 runs. Shaded regions show the 95% confidence intervals. Right: Each data point is plotted, as well as the mean values which are marked with dark red. Horizontal lines indicate statistically different results. ***: $P < 0.001$, **: $P < 0.005$, *: $P < 0.05$ (Color figure online)

different settings by varying the environment, Walker-v0 and BridgeWalker-v0, and the morphology space, $(h, w) \in \{(5, 5), (7, 7)\}$. Each setting is named as {Environment name}{(h,w)}, e.g. **W5** for the setting **W**alker-v0, (**5**, 5). For each setting, we repeat the experiment 10 times and use the Wilcoxon Rank Sum test [58] to report P-values. Each run consists of a population size of 16 individuals evolved for 10000 generations with the AFPO algorithm [47].

Figure 1 left, illustrates the fitness over evolutionary time plots for each experiment. Solid lines show the fitness of the best solution found at each generation, averaged across 10 runs. Shaded regions show the 95% bootstrapped confidence interval. Additionally, Fig. 1 right illustrates the distribution of the performance of the final solutions. The results demonstrate that the populations with fixed controllers find better solutions faster (all $P < 0.05$ for achieving 85% of final performance) and find significantly better solutions (all $P < 0.005$). These results are consistent across different environments and morphology spaces. While individuals with fixed controllers are, admittedly, incapable of observing their environment, hence unlikely to scale to more complex problems, *fixed controllers nevertheless provide a strong baseline for performance for simple tasks such as locomotion, where a lot of research is still being done.*

The success of fixed controller might seem interesting, yet it is a well-established phenomenon that it is difficult to optimize body plan and control together [3, 5, 20, 40]. Having two interdependent parts of a solution creates a challenging optimization problem where the parts become more and more specialized for each other as the optimization takes place, making them unamenable to change without breaking the solution's performance – fragile co-adaptation. This is especially true for morphology, as it is the interface between the controller and environment and thus changing morphology also scrambles the controller. It makes it harder to search the morphology space, resulting in premature convergence of morphology [3,20]. On the other hand, the lack of parameters for the fixed controller turns the co-optimization of the morphology and control problem into only the optimization of the morphology, eliminating the issue of fragile co-adaptation. The results demonstrating superior performance for fixed controllers suggest that *fixed controller doesn't suffer from premature convergence and allows optimization process to find body plans that outsource complexity to the dynamics of the system, an example of morphological computation* [42].

Lastly, we note that the fixed controller consistently outperforms the more complex learnable controller in the BridgeWalker-v0 environment in two different morphology spaces. In light of these results, *we conjecture that the locomotion over uneven surface doesn't present fundamentally harder/different optimization problem than locomotion on flat surface for brain-body co-optimization.*

4   Analysis

Our experiments demonstrate the inferior performance of co-optimization with learnable controllers compared to morphology optimization with fixed controllers. In this section, we perform a series of analyses to understand how the differences in controllers affect the co-optimization. Since we expect learnable controllers to be capable of learning to output the simple behavior that the fixed controllers exhibit, the differences in performance should arise from the different controllers' effects on the search over the morphology space. As prior work strongly emphasizes the issue of premature convergence as the major hurdle in the brain-body co-optimization [3,20,40], we specifically focus on this issue in our analysis.

Qualitative Analysis of Body Plans. As we discussed earlier, one of the biggest challenges of brain-body co-optimization is the premature convergence of the morphology [3,20,40]. While optimizing a controller for a given body plan is relatively easy, search over the morphology space often prematurely converges to a local optimum. To compare experimented controllers in terms of their effect on the search over the morphology space, we start by looking at the body plans of the best solutions found at each run, i.e. run champions, under each setting.

First, we try to measure the run champions' convergence to a particular body plan, which indicates a better search over the morphology space as we expect a successful search to be able to exploit high-performing regions of the space across runs [3]. To this end, we perform t-SNE dimensionality reduction [34] of the optimized morphologies. We visualize the clusters in Fig. 2 and measure the average intra-cluster distance, to quantify the variation in found morphologies for a given controller paradigm. We find that run champions with fixed controllers

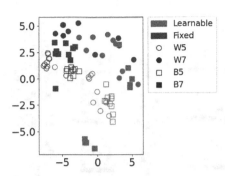 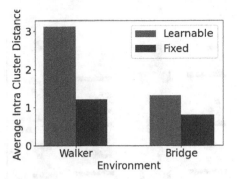

Fig. 2. t-SNE plots of the run champions' body plans and average intra-cluster distances in the embedding space. The populations with the fixed controller converge better to a body plan across different runs and morphology spaces.

converge to more similar solutions across different runs and morphology spaces in two different environments. We also note that the learnable run champions show less diversity in the BridgeWalker-v0 environment, showing that the environment also plays a role in shaping the fitness landscape and affecting which body plans can be easily found during the co-optimization process.

To qualitatively analyze the robot body plans and gaits, we present Fig. 3, where we show the body plans of the run champions, and Fig. 4, where we display 15 snapshots during the lifetime of select individuals. In our experimental setup, fixed controllers (Fig. 4b) consistently find similar body plans with near identical gaits. We do not observe this level of converge across runs that co-optimize morphology with a learnable controller (Fig. 4a). This trend holds across all experimental settings.

Quantitative Analysis of Body Plans. Our qualitative analysis of body plans shows that champions with the learnable controller show more diversity while fixed run champions are grouped closer in the morphology space. The diversity of solutions found with learnable controllers can be interpreted as a positive feature. However, the literature strongly suggests that it is a sign of poor search over the morphology space [3]. Although it is possible that the search space contains many distinct near-optimal solutions, the high-performing body plans of fixed run champions imply the existence of better-performing body plans that aren't discovered in runs with the learnable controller.

Therefore in this section, we try to definitively answer our main question: have learnable controllers failed to find high-performing body plans that were discovered by fixed controllers? To answer this question, we take the morphologies of the run champions found in runs with fixed controllers and no co-optimization, then optimize a learnable controller for each body plan. We suggest that by pairing the same learned closed-loop controller from the co-optimization paradigm to the morphologies found when optimized with the fixed controller, we will be able to fairly compare how well each optimization setup is able to find

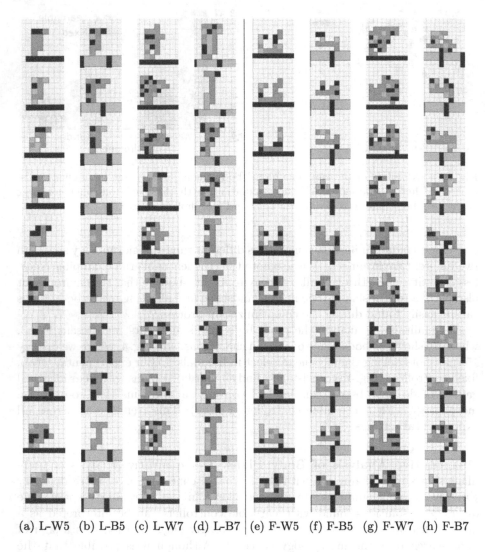

(a) L-W5   (b) L-B5   (c) L-W7   (d) L-B7 | (e) F-W5   (f) F-B5   (g) F-W7   (h) F-B7

Fig. 3. Morphologies of the run champions under each experimental setting. The left half of the figure shows the run champions with learnable controllers, and the right half shows the ones with fixed controllers. We see diverse body plans evolve with the learnable controller, especially in (a). On the other hand, we see the same form of the mostly active bottom with two vertical apparatus at the front and back, resembling a head and a tail, evolved with the fixed controller in (e). For the BridgeWalker-v0 environment, individuals with learnable controllers show similar body plan features in (b) and (d), e.g. upright posture, r or T shape, individuals with fixed controller usually consists of a thin, horizontal body with active materials and forward apparatus resembling a leg in (f–h).

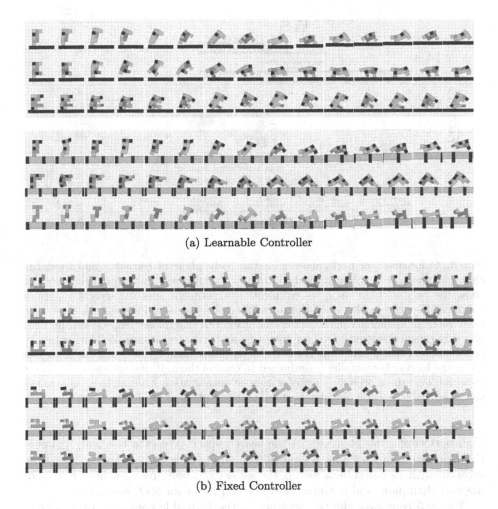

(a) Learnable Controller

(b) Fixed Controller

Fig. 4. Timelapse images of three run champions' behavior from each of the two tasks of the (5, 5) morphology space, co-optimized brain and body (a) and morphology-only optimization with fixed control (b). Learnable controllers demonstrate more diversity, especially in the Walker-v0 environment. Most of the individuals with learnable controllers (a) find a bipedal (rows 1, 2, 4–6) or monopedal gait (row 3). Fixed controllers (b) show one common behavior on flat ground (rows 1–3), consisting of an active bottom with two vertical apparatus at the front and back, resembling a head and a tail. When the muscle at the bottom of the robot contracts, vertical apparatuses help the bottom part form an arch, creating front and back legs to locomote. In the bridge environment (rows 4–6) often includes a thin, horizontal body with active materials, a forward apparatus resembling a leg, and some upper body (presumably used for balancing). Individuals with this body plan throw themselves forward by actuating their muscles in phase and pulling themselves forward with their forward apparatus, achieving bipedal locomotion.

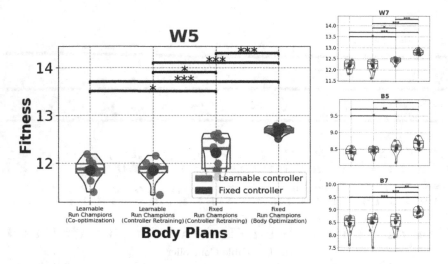

Fig. 5. Comparison of performances on the body plans of learnable and fixed run champions with learnable and fixed controllers. The first and last columns in each plot show the results of the main co-optimization experiment, and the second and third columns show the results when the learnable controller is optimized from scratch on body plans of learnable run champions and fixed run champions, respectively. In 3 out of 4 experimental settings, the body plans found by the fixed controller during the co-optimization experiments outperform the body plans found by the learnable controller when the learnable controller is optimized to control them. It demonstrates the failure of search over the morphology space during co-optimization with learnable controllers since they failed to discover high-performing body plans of fixed champions.

high-performing morphologies in the search space. We take the body plans of the run champions with fixed controllers and start new evolutionary runs from scratch, with 16 individuals having the same fixed body plan as the corresponding run champion, and optimize only the controller for 5000 generations.

Figure 5 compares the performance of the learnable controller on the body plans of run champions found during co-optimization with learnable and morphology/only optimization with fixed controllers. For comparison, we also show the original performances of the fixed run champions with the fixed controller. In 3 out of 4 experimental settings (W5, W7, and B5; all $P < 0.05$), the learnable controller achieves significantly better performance on the body plans of fixed champions, which definitively demonstrates the existence of better body plans than the ones found during co-optimization with the learnable controller. *To the best of our knowledge, this is the most concrete example to date showing the failure of brain-body co-optimization to find specific and demonstratedly-reachable high-performing regions of the morphology space.*

It is also possible that simply the act of re-training the controller once the body plan has already been found and fixed produces a much different optimization trajectory than co-optimizing both at once. To account for this we also take the run champions found when co-optimizing morphology with a learnable controller, freeze the resulting morphology and re-train a learnable controller on it from scratch (emulating the setup described above). Across all conditions, this resulted in statistically indistinguishable performance from the original co-optimized robots (Fig. 5, first versus second columns, all $P > 0.44$), demonstrating that the controllers found during co-optimization were well already converged to their current body plan. This finding also provides further evidence consistent with the notion that the struggle to co-optimize brain-body systems is due to sub-par morphologies more so than sub-par controllers.

For completeness, we also measured the performance of learnable run champions' body plans when they were controlled by the fixed controller. As expected, they performed very poorly (mean performance in W5: 4.88, W7: 6.31, B5: 2.64, B7: 2.47) and most solutions couldn't even reach the end of the environment. This suggests that the fixed controllers are not inherently better than the learnable controllers, but benefit from enabling search to discover more effective body plans, and especially body plans that are well adapted to the given control policy.

Lastly, contrary to what we expect, we observe that the learnable controller, when it is trained on fixed run champions' body plans, doesn't always achieve the same level of performance as fixed controllers do (Fig. 5, third versus last columns, $P < 0.05$ in W5, W7, and B7), which we interpret as the challenge of achieving very simple behaviors with complex neural network controllers when they are optimized through fitness alone. Future work could consider experimenting with more complex tasks and environments to understand how the co-optimization process unfolds when the complexity of the task/environment better matches the complexity of the controller.

5   Discussion

Our co-optimization experiment and the following analysis produced a number of important results that can be seen in Fig. 5. Most importantly, we have shown that optimizing learnable controllers on the high-performing body plans found with fixed controllers, resulted in better solutions compared to the solutions found with learnable controllers during co-optimization in most experimental settings (first vs middle columns, $P < 0.05$ in 3 out of 4), which clearly demonstrates the failure of the search over the morphology space with learnable controllers during co-optimization, providing more evidence for the premature convergence of morphology as the main challenge of brain-body co-optimization.

While Cheney et al. [3] provides an embodied cognition perspective into this phenomenon, here we try to understand it from an optimization perspective. The standard intuition conceives a solution space where each solution has a single corresponding fitness creating a landscape over the solution space. However rugged it may be, we hope solutions with similar fitness values are grouped together in the solution space at some resolution, which allows us to search the space with heuristic search methods. This is certainly applicable to the brain-body co-optimization problem as well, where the brain and body together create the solution space. But instead of visualizing a standard landscape with one fitness value per brain-body pair, we could alternatively conceptualize two solution domains: one a fitness landscape over controllers and another a fitness landscape over body plans. In doing so, we notice a peculiarity about brain-body solution space. While small changes in one part of the solution, that being the brain, can result in solutions with similar fitness and thus produce a typical and relatively smooth fitness landscape, changes to a robot's body almost always result in a severe fitness drop [40]. In light of this insight, we advocate for a new morphology-centric way of conceptualizing the fitness landscape that the co-optimization process runs on. *Rather than imagining a fixed fitness value per each solution candidate, we can conceptualize the fitness landscape as a surface over the morphology space with a range of possible fitness values per body plan, where the actual fitness depends on the controller.*

This new view allows us to understand the premature convergence of morphology from a search perspective. Body plans that we found early in the co-optimization process accrue more controller optimization steps, climbing up to their maximum fitness value – an example of a first-mover advantage [31]. Conversely, body plans we found later on during the search via mutation, however promising these body plans may be, are at a disadvantage: they can't compete with first-movers until a controller is optimized for them to provide a better assessment of their maximum fitness, but typically can't survive long enough to accrue control optimization if they can't compete with first movers initially.

Indeed, Cheney et al. [3] investigate this by designing a smaller search space for the control which reduces the range of possible fitness values per body plan, effectively limiting the first-movers advantage, and showing that the premature convergence of morphology disappears. We take this even further by using a fixed controller which collapses the range of possible fitness values per body plan into a single value, turning the fitness landscape into a familiar format that we can search effectively. The success of the fixed controller shows that our algorithms are adequate in searching the standard landscape, but confronted with the peculiar co-optimization landscape that is unforgiving to changes in the body plan, they fail and get stuck in local optima.

Luckily, this new understanding of the fitness landscape allows us to put recently proposed solutions into perspective: *if the search over the morphology space is failing due to our inability to effectively measure the true potential of body plans, then (1) we could give our algorithms more resources to assess new body plans [5], and/or (2) we could try to have a better initial assessment for new body plans by alleviating the performance decline during the morphological search by using different genetic representations for coordinated changes [54,57] or by using controllers that are robust to changes in the morphology [40].* These two ways are orthogonal to each other and we believe further research on these avenues holds the potential for unlocking successful brain-body co-optimization.

It also enables us to draw connections to other fields. For example, considering the optimization of the controller for a given morphology to be an approximation of the true potential (i.e. maximum) fitness of that morphology. Thus our framework connects brain-body co-optimization with a wealth of literature on surrogate proxy fitness models [19,50], on approaches to optimization under uncertainty such as for the multi-arm bandit problems [1,29]. In robot brain-body co-optimization the computation expense of fitness evaluations makes it unreasonable to train individual controllers from scratch to ideally and independently determine the fitness of each morphology. This is also the case for training the weights for a deep neural network architecture, and this framing draws analogies to the notion of sharing controllers for all possible body-plans/architectures in the fitness landscape, deemed supernet weight-sharing algorithms in the Neural Architecture Search literature [32,45]. It also suggests the importance of methods to enable effective transfer of controllers from one morphology to another when sharing these fitness approximators across disparate body plans via approaches like meta-learning, transfer learning, and few-shot learning [12,40,60,61].

Moreover, the success of body optimization with fixed controllers (Fig. 5, first vs last columns, all $P < 0.005$) brings the formulation of "brain" and "body" into question. Formulating the control of the robot in ways that it assigns a single or smaller range of fitness values unlocks the body plan optimization by collapsing the fitness landscape and limiting the first-movers advantage. Solutions where robots' materials react to their environment in simpler ways and complex behavior emerges by exploiting the dynamics of the body and interaction with the environment, such as [4,6,8–10], can be further investigated. Especially, scaling this methodology to solve more complex tasks that would normally assumed to require complex closed-loop decision making should be investigated.

6   Conclusion

We have compared learnable neural network controllers and heuristically designed fixed controllers in the problem of co-optimization of brain and body (while for the latter, it is just the optimization of the body) for the locomotion task in multiple environments and morphology spaces that vary in complexity. Our experiments demonstrated the existence of high-performing regions that weren't found during the co-optimization process. Based on our analysis with the fixed controller, we developed a new morphology-centric understanding of the fitness landscape which explains the premature convergence from a search perspective and clarifies previously proposed solutions. We hope this new framework will help us systematically tackle the challenges of brain-body co-optimization.

Acknowledgements. This material is based upon work supported by the National Science Foundation under Grant No. 2008413, 2239691, 2218063. Computations were performed on the Vermont Advanced Computing Core supported in part by NSF Award No. OAC-1827314.

References

1. Agrawal, S., Goyal, N.: Analysis of Thompson sampling for the multi-armed bandit problem. In: Conference on Learning Theory, p. 39-1. JMLR Workshop and Conference Proceedings (2012)

2. Bhatia, J., Jackson, H., Tian, Y., Xu, J., Matusik, W.: Evolution gym: a large-scale benchmark for evolving soft robots. Adv. Neural. Inf. Process. Syst. **34**, 2201–2214 (2021)
3. Cheney, N., Bongard, J., Sunspiral, V., Lipson, H.: On the difficulty of co-optimizing morphology and control in evolved virtual creatures. In: Proceedings of the Artificial Life Conference 2016, pp. 226–233. MIT Press, Cancun (2016). https://doi.org/10.7551/978-0-262-33936-0-ch042
4. Cheney, N., Bongard, J., Lipson, H.: Evolving soft robots in tight spaces. In: Proceedings of the 2015 Annual Conference on Genetic and Evolutionary Computation, pp. 935–942. ACM, Madrid (2015). https://doi.org/10.1145/2739480.2754662
5. Cheney, N., Bongard, J., SunSpiral, V., Lipson, H.: Scalable co-optimization of morphology and control in embodied machines. J. R. Soc. Interface **15**(143), 20170937 (2018). https://doi.org/10.1098/rsif.2017.0937
6. Cheney, N., MacCurdy, R., Clune, J., Lipson, H.: Unshackling evolution: evolving soft robots with multiple materials and a powerful generative encoding, p. 8 (2014)
7. Corucci, F., Cheney, N., Kriegman, S., Bongard, J., Laschi, C.: Evolutionary developmental soft robotics as a framework to study intelligence and adaptive behavior in animals and plants. Front. Robot. AI **4**, 34 (2017). https://doi.org/10.3389/frobt.2017.00034
8. Corucci, F., Cheney, N., Lipson, H., Laschi, C., Bongard, J.: Material properties affect evolutions ability to exploit morphological computation in growing soft-bodied creatures. In: Proceedings of the Artificial Life Conference 2016, pp. 234–241. MIT Press, Cancun (2016). https://doi.org/10.7551/978-0-262-33936-0-ch043
9. Corucci, F., Cheney, N., Lipson, H., Laschi, C., Bongard, J.C.: Evolving swimming soft-bodied creatures, p. 2 (2016)
10. Cheney, N., Clune, J., Lipson, H.: Evolved electrophysiological soft robots. In: Artificial Life 14: Proceedings of the Fourteenth International Conference on the Synthesis and Simulation of Living Systems, pp. 222–229. The MIT Press (2014). https://doi.org/10.7551/978-0-262-32621-6-ch037
11. Ferigo, A., Iacca, G., Medvet, E., Pigozzi, F.: Evolving Hebbian learning rules in voxel-based soft robots. Preprint (2021). https://doi.org/10.36227/techrxiv.17091218.v1
12. Finn, C., Abbeel, P., Levine, S.: Model-agnostic meta-learning for fast adaptation of deep networks. In: International Conference on Machine Learning, pp. 1126–1135. PMLR (2017)
13. Hiller, J., Lipson, H.: Automatic design and manufacture of soft robots. IEEE Trans. Rob. **28**(2), 457–466 (2012). https://doi.org/10.1109/TRO.2011.2172702
14. Hiller, J., Lipson, H.: Dynamic simulation of soft multimaterial 3D-printed objects. Soft Rob. **1**(1), 88–101 (2014). https://doi.org/10.1089/soro.2013.0010
15. Hiller, J.D., Lipson, H.: Evolving amorphous robots. In: Alife, pp. 717–724. Citeseer (2010)
16. Horibe, K., Walker, K., Berg Palm, R., Sudhakaran, S., Risi, S.: Severe damage recovery in evolving soft robots through differentiable programming. Genet. Program Evolvable Mach. **23**(3), 405–426 (2022)
17. Horibe, K., Walker, K., Risi, S.: Regenerating soft robots through neural cellular automata. In: Hu, T., Lourenço, N., Medvet, E. (eds.) EuroGP 2021. LNCS, vol. 12691, pp. 36–50. Springer, Cham (2021). https://doi.org/10.1007/978-3-030-72812-0_3
18. Huang, W., Mordatch, I., Pathak, D.: One policy to control them all: shared modular policies for agent-agnostic control. In: International Conference on Machine Learning, pp. 4455–4464. PMLR (2020)

19. Jin, Y.: Surrogate-assisted evolutionary computation: recent advances and future challenges. Swarm Evol. Comput. **1**(2), 61–70 (2011)
20. Joachimczak, M., Suzuki, R., Arita, T.: Artificial metamorphosis: evolutionary design of transforming, soft-bodied robots. Artif. Life **22**(3), 271–298 (2016). https://doi.org/10.1162/ARTL_a_00207
21. Kim, S., Laschi, C., Trimmer, B.: Soft robotics: a bioinspired evolution in robotics. Trends Biotechnol. **31**(5), 287–294 (2013)
22. Kriegman, S., Blackiston, D., Levin, M., Bongard, J.: A scalable pipeline for designing reconfigurable organisms. Proc. Natl. Acad. Sci. **117**(4), 1853–1859 (2020). https://doi.org/10.1073/pnas.1910837117
23. Kriegman, S., Blackiston, D., Levin, M., Bongard, J.: Kinematic self-replication in reconfigurable organisms. Proc. Natl. Acad. Sci. **118**(49), e2112672118 (2021). https://doi.org/10.1073/pnas.2112672118
24. Kriegman, S., Cheney, N., Bongard, J.: How morphological development can guide evolution. Sci. Rep. **8**(1), 13934 (2018). https://doi.org/10.1038/s41598-018-31868-7
25. Kriegman, S., Cheney, N., Corucci, F., Bongard, J.C.: Interoceptive robustness through environment-mediated morphological development. In: Proceedings of the Genetic and Evolutionary Computation Conference, pp. 109–116. ACM, Kyoto (2018). https://doi.org/10.1145/3205455.3205529
26. Kriegman, S., et al.: Scale invariant robot behavior with fractals. In: Robotics: Science and Systems XVII. Robotics: Science and Systems Foundation (2021). https://doi.org/10.15607/RSS.2021.XVII.059
27. Kriegman, S., et al.: Scalable sim-to-real transfer of soft robot designs. In: 2020 3rd IEEE International Conference on Soft Robotics (RoboSoft), pp. 359–366. IEEE (2020)
28. Kriegman, S., Walker, S., Shah, D.S., Levin, M., Kramer-Bottiglio, R., Bongard, J.: Automated shapeshifting for function recovery in damaged robots. In: Robotics: Science and Systems XV. Robotics: Science and Systems Foundation (2019). https://doi.org/10.15607/RSS.2019.XV.028
29. Kuleshov, V., Precup, D.: Algorithms for multi-armed bandit problems. arXiv preprint arXiv:1402.6028 (2014)
30. Lehman, J., Stanley, K.O.: Evolving a diversity of virtual creatures through novelty search and local competition. In: Proceedings of the 13th Annual Conference on Genetic and Evolutionary Computation - GECCO 2011, p. 211. ACM Press, Dublin (2011). https://doi.org/10.1145/2001576.2001606
31. Lieberman, M.B., Montgomery, D.B.: First-mover advantages. Strategic Manage. J. **9**(S1), 41–58 (1988)
32. Liu, H., Simonyan, K., Yang, Y.: DARTS: differentiable architecture search. In: International Conference on Learning Representations (2019). https://openreview.net/forum?id=S1eYHoC5FX
33. Liu, S., Matthews, D., Kriegman, S., Bongard, J.: Voxcraft-sim, a GPU-accelerated voxel-based physics engine (2020)
34. Van der Maaten, L., Hinton, G.: Visualizing data using t-SNE. J. Mach. Learn. Res. **9**(11) (2008)
35. Marzougui, D., Biondina, M.: A comparative analysis on genome pleiotropy for evolved soft robots, p. 4 (2022)
36. Medvet, E., Bartoli, A., De Lorenzo, A., Fidel, G.: Evolution of distributed neural controllers for voxel-based soft robots. In: Proceedings of the 2020 Genetic and Evolutionary Computation Conference, pp. 112–120. ACM, Cancún (2020). https://doi.org/10.1145/3377930.3390173

37. Medvet, E., Bartoli, A., De Lorenzo, A., Seriani, S.: 2D-VSR-sim: a simulation tool for the optimization of 2-D voxel-based soft robots. SoftwareX **12**, 100573 (2020). https://doi.org/10.1016/j.softx.2020.100573
38. Medvet, E., Bartoli, A., Pigozzi, F., Rochelli, M.: Biodiversity in evolved voxel-based soft robots. In: Proceedings of the Genetic and Evolutionary Computation Conference, pp. 129–137. ACM, Lille (2021). https://doi.org/10.1145/3449639.3459315
39. Medvet, E., Nadizar, G., Pigozzi, F.: On the impact of body material properties on neuroevolutionfor embodied agents: the case of voxel-based soft robots, p. 9 (2022)
40. Mertan, A., Cheney, N.: Modular controllers facilitate the co-optimization of morphology and control in soft robots. In: Proceedings of the Genetic and Evolutionary Computation Conference, pp. 174–183. ACM, Lisbon (2023). https://doi.org/10.1145/3583131.3590416
41. Pfeifer, R., Bongard, J.: How the body shapes the way we think: a new view of intelligence (2006)
42. Pfeifer, R., Iida, F.: Morphological computation: connecting body, brain, and environment, p. 5 (2009)
43. Pigozzi, F., Tang, Y., Medvet, E., Ha, D.: Evolving modular soft robots without explicit inter-module communication using local self-attention. In: Proceedings of the Genetic and Evolutionary Computation Conference, pp. 148–157 (2022)
44. Pontes-Filho, S., Walker, K., Najarro, E., Nichele, S., Risi, S.: A single neural cellular automaton for body-brain co-evolution, p. 4 (2022)
45. Ren, P., et al.: A comprehensive survey of neural architecture search: challenges and solutions. ACM Comput. Surv. **54**(4) (2021). https://doi.org/10.1145/3447582
46. Rus, D., Tolley, M.T.: Design, fabrication and control of soft robots. Nature **521**(7553), 467–475 (2015)
47. Schmidt, M.D., Lipson, H.: Age-fitness pareto optimization, p. 2 (2010)
48. Shah, D.S., Powers, J.P., Tilton, L.G., Kriegman, S., Bongard, J., Kramer-Bottiglio, R.: A soft robot that adapts to environments through shape change. Nat. Mach. Intell. **3**(1), 51–59 (2021)
49. Shepherd, R.F., et al.: Multigait soft robot. Proc. Natl. Acad. Sci. **108**(51), 20400–20403 (2011)
50. Shi, L., Rasheed, K.: A survey of fitness approximation methods applied in evolutionary algorithms. In: Tenne, Y., Goh, C.-K. (eds.) Computational Intelligence in Expensive Optimization Problems. ALO, vol. 2, pp. 3–28. Springer, Heidelberg (2010). https://doi.org/10.1007/978-3-642-10701-6_1
51. Sims, K.: Evolving virtual creatures. In: Proceedings of the 21st annual conference on Computer graphics and interactive techniques - SIGGRAPH 1994, pp. 15–22. ACM Press, Not Known (1994). https://doi.org/10.1145/192161.192167
52. Talamini, J., Medvet, E., Bartoli, A., Lorenzo, A.D.: Evolutionary synthesis of sensing controllers for voxel-based soft robots, p. 8 (2019)
53. Talamini, J., Medvet, E., Nichele, S.: Criticality-driven evolution of adaptable morphologies of voxel-based soft-robots. Front. Robot. AI **8** (2021). https://www.frontiersin.org/article/10.3389/frobt.2021.673156
54. Tanaka, F., Aranha, C.: Co-evolving morphology and control of soft robots using a single genome. In: 2022 IEEE Symposium Series on Computational Intelligence (SSCI), pp. 1235–1242. IEEE (2022)
55. Trimmer, B.A.: New challenges in biorobotics: incorporating soft tissue into control systems. Appl. Bionics Biomech. **5**(3), 119–126 (2008). https://doi.org/10.1080/11762320802617255

56. Veenstra, F., Glette, K.: How different encodings affect performance and diversification when evolving the morphology and control of 2D virtual creatures. In: The 2020 Conference on Artificial Life, pp. 592–601. MIT Press (2020). https://doi.org/10.1162/isal_a_00295
57. Veenstra, F., Olsen, M.H., Glette, K.: Effects of encodings and quality-diversity on evolving 2D virtual creatures, p. 4 (2022)
58. Wilcoxon, F., Wilcox, R.A.: Some rapid approximate statistical procedures. (No Title) (1964)
59. Yim, M., et al.: Modular self-reconfigurable robot systems [grand challenges of robotics]. IEEE Robot. Autom. Mag. **14**(1), 43–52 (2007). https://doi.org/10.1109/MRA.2007.339623
60. Yosinski, J., Clune, J., Bengio, Y., Lipson, H.: How transferable are features in deep neural networks? In: Adv. Neural Inf. Process. Syst. **27** (2014)
61. Zhao, Y., Wang, L., Tian, Y., Fonseca, R., Guo, T.: Few-shot neural architecture search. In: International Conference on Machine Learning, pp. 12707–12718. PMLR (2021)

Grammar-Based Evolution
of Polyominoes

Jessica Mégane[1] , Eric Medvet[2(✉)] , Nuno Lourenço[1] ,
and Penousal Machado[1]

[1] University of Coimbra, CISUC/LASI - Centre for Informatics and Systems of the
University of Coimbra, Department of Informatics Engineering, Coimbra, Portugal
{jessicac,naml,machado}@dei.uc.pt
[2] Department of Engineering and Architecture, University of Trieste, Trieste, Italy
emedvet@units.it

Abstract. Languages that describe two-dimensional (2-D) structures
have emerged as powerful tools in various fields, encompassing pattern
recognition and image processing, as well as modeling physical and chemical phenomena. One kind of two-dimensional structures is given by
labeled polyominoes, i.e., geometric shapes composed of connected unit
squares represented in a 2-D grid. In this paper, we present (a) a novel
approach, based on grammars, for describing sets of labeled polyominoes that meet some predefined requirements and (b) an algorithm to
develop labeled polyominoes using the grammar. We show that the two
components can be used for solving optimization problems in the space
of labeled polyominoes, similarly to what happens for strings in grammatical evolution (and its later variants). We characterize our algorithm
for developing polyominoes in terms of representation-related metrics
(namely, validity, redundancy, and locality), also by comparing different
representations. We experimentally validate our proposal using a simple
evolutionary algorithm on a few case studies where the goal is to obtain a
target polyomino: we show that it is possible to enforce hard constraints
in the search space of polyominoes, using a grammar, while performing
the evolutionary search.

Keywords: polyomino · grammar · representation · 2-D patterns

1 Introduction

Two-dimensional (2-D) languages have emerged as powerful tools in various
fields, initially motivated by problems in pattern recognition and image processing [4,6,14]. These languages generate 2-D objects and can be designed to generate either a simple rectangle or a more intricate shape such as a decagon. One
shape that has garnered significant interest due to its unique properties [1,5,13]
and wide-ranging applicability [8,15,21,25,26,26,28,29] is the polyomino.

Polyominoes [7] are geometric shapes composed of connected unit squares,
forming a finite set of cells within a 2-D grid. These shapes are also commonly

referred to as lattice animals in the physical [8] and chemical [26] fields, where they have been popularly used to model branched polymers, molecules and percolation processes, providing valuable insights into the behavior of these complex systems [3,36]. The field of combinatorial optimization and mathematics has extensively explored polyominoes, due to their rich mathematical properties [1,5,13]. Similarly, the theoretical formal language domain has been inspired by their practical usages.

In these fields, it is often crucial to find one or more polyominoes that maximize specific objectives while satisfying predefined structural requirements. Two powerful mechanisms for describing and generating 2-D shapes are grammars [15,25,27,29,37].

Grammars define a language with a set of rules that can impose certain constrains. These grammars can be designed either as one-dimensional encodings of two-dimensional structures [37] or be two-dimensional structured representations [27,29]. To the best of our knowledge, there exists only a single approach that utilizes grammars to define polyominoes constraints and then attempts to optimize them [37], namely, for finding an assembly of identical polyominoes. While this represents the only usage of a grammar to define polyominoes in the literature, numerous examples exist where grammars are employed to generate 2-D pictures [12,14,18,34], finding practical applications in popular tasks such as mathematical formula recognition [15,25,29].

Polyominoes can be further enhanced by assigning labels to individual cells, providing additional information for each cell within the structure. In this paper, we present a novel approach for generating labeled polyominoes that meet some structural requirements defined in a formal way: for this purpose, we (a) define the concept of polyomino context-free grammars (PoCFGs) as an extension of context-free grammars (CFGs) and (b) propose a development algorithm that can be used for generating a polyomino adhering to a PoCFGs. Our proposed algorithm constructs these polyomino structures with precise control of the shape of the polyomino and labeling of its cells. When used inside an evolutionary algorithm (EA), it allows solving optimization problems over the space of labeled polyominoes adhering to a given PoCFG. This process only requires the provision of a grammar and a fitness function to the selected EA; notably, it does not require users to provide variation operators that guarantee that varied polyominoes will still adhere to the PoCFG—i.e., operators with the closure property. This fact greatly increases the applicability of our approach, lowering the barrier to polyominoes optimization, much like grammatical evolution (GE) [31] did with regular languages.

Since our algorithm is greatly agnostic with respect to the genotypic representation employed by the EA, we compare different representations in terms of representation-related metrics (validity, redundancy, and locality). Moreover, to showcase the effectiveness of our approach, we evolve some polyominoes in a few case studies where the goal is to evolve a polyomino adhering to a grammar that is as much as possible similar to a pre-defined target polyomino. We show experimentally that evolutionary optimization does work, giving polyomi-

noes that are more and more similar to the target one while all adhering to the
provided grammar, i.e., meeting the user-defined constraints.

2   Our Proposal: Describing and Evolving Polyominoes

2.1   Labeled Polyomino

A *polyomino* is a 2-D geometric figure formed by one or more squares (or *cells*)
joined together along their sides. A *labeled polyomino* defined over an alphabet A
is a polyomino in which each cell is associated with exactly one symbol (or *label*)
$a \in A$. For brevity, from now on we will write simply polyomino for referring to
labeled polyominoes. We denote by $\mathcal{P}_A$ the set of all the polyominoes defined
over A.

By assigning a coordinate $(x_0, y_0) \in \mathbb{Z}^2$ to one of the cells of a polyomino p,
we denote by $p_{x,y} \in A$ the label of the cell of p displaced by $x - x_0$ cells along
the x-axis and $y - y_0$ cells along the y-axis with respect to the cell at (x_0, y_0).
We write $p_{x,y} = \varnothing$ if there is no cell at (x, y) in p.

A *referenced polyomino* is a polyomino in which one cell is identified as *reference cell*. By convention, we assume that the reference cell is assigned to the
coordinate $(0, 0)$.

2.2   Polyomino Context-Free Grammar (PoCFG)

A polyomino context-free grammar (PoCFG) $\mathcal{G}$ is a tuple $\mathcal{G} = (N, T, n_1, \mathcal{R})$
where N is a finite set of *non-terminal* symbols, T is a finite set of *terminal*
symbols, with $N \cap T = \emptyset$, $n_1 \in N$ is the starting symbol (or *axiom*), and $\mathcal{R}$
is a finite set of *production rules*. A production rule is a pair composed of a
non-terminal symbol (the left-hand-side of the rule) and a referenced polyomino
defined over the alphabet $N \cup T$ (the right-hand-side).

Similarly to the case of CFGs for strings, we represent a PoCFG with a
compact notation which resembles the Backus-Naur form (BNF). In BNF rules
are grouped together by their non-terminal symbol and the first rule is the one
for the starting symbol n_1. Figure 2 shows an example of five PoCFGs in BNF
(the ones used in our experiments).

As for the case of CFGs for strings, a PoCFG $\mathcal{G} = (N, T, n_1, \mathcal{R})$ is a compact
way for defining a (possibly infinite) set of polyominoes defined over T; we denote
by $\mathcal{P}_\mathcal{G} \subseteq \mathcal{P}_T$ the set of polyominoes defined by $\mathcal{G}$. In the next section, we describe
a constructive process that allows to obtain one $p \in \mathcal{P}_\mathcal{G}$. Note that the problem
of deciding whether a given polyomino p belongs to $\mathcal{P}_\mathcal{G}$ is beyond the scope of
this paper—for CFGs that meet some requirements, this problem is solvable for
the case of strings [32].

2.3   PoCFG-Based Development Algorithm

We propose a *development algorithm* for obtaining a polyomino $p \in \mathcal{P}_\mathcal{G}$ for a
PoCFG $\mathcal{G}$. We call it development algorithm because it iteratively modifies a

polyomino by either adding new cells or modifying existing ones according to the production rules of G and starting from a single cell polyomino given by the axiom—from this point of view, it resembles a developmental process.

Design Principles. Since the eventual usage of this algorithm is within the process of the evolutionary optimization over $\mathcal{P}_G$, we designed it to receive as input a source of information that is used for choosing which production rules to apply. Consistently with the intended usage, we call this input the *genotype* g. The development algorithm hence maps a genotype g to a polyomino p.

We designed our algorithm to be largely agnostic to the kind of genotype being fed as input, i.e., to $G \ni g$. We achieved this goal by making the algorithm modular, i.e., by decoupling the part where a suitable production rule is chosen using g from the rest of the algorithm, namely from the part in which the chosen rule is used and the part in which suitable rules are identified. Indeed, in Sect. 3.2 we experimentally compare realizations of the development algorithm where the genotype is a bit- or an integer-string or a more complex data structure.

Working. Algorithm 1 presents our development algorithm in the form of pseudocode. It takes as input the genotype $g \in G$ and the PoCFG G and, as parameters, a sorting criterion c and an overwriting flag o; it returns either a polyomino $p \in \mathcal{P}_G$ or $\varnothing$ if it is not possible to develop a polyomino with the given inputs and parameters.

The algorithm works as follows. First, it sets (line 2) p to the one-cell polyomino with the only cell being labeled with the axiom n_1. Then, it iteratively modifies p according to these steps: (i) it finds (line 5) all the cells in p that are labeled with a non-terminal symbols in N, i.e., those which can be replaced according to a rule in $\mathcal{R}$; (ii) it chooses (line 9) one cell $(x^\star, y^\star)$ to be the target of the replacement using the sorting criterion c; (iii) it chooses (line 11), based on the genotype g and the state s (initialized to $\varnothing$, see Sect. 2.3), one rule to apply among the ones suitable for the cell at $(x^\star, y^\star)$; (iv) finally, if possible, it performs (line 18) the replacement in p according to the chosen rule. The last step, i.e., the actual modification of p, consists in "putting" the referenced polyomino p' "over" p with the reference cell of p' placed at $(x^\star, y^\star)$ in p. The iterations stop if (a) no more cells labeled with a non-terminal symbol are present in the polyomino or (b) some of the steps cannot be performed (see below).

Step (ii) above (FIRST() in Algorithm 1) corresponds in selecting one nonterminal cell to be the target of the replacement—in general, there can be more than one non-terminal cell in the polyomino being developed at some point of the mapping. The way this choice is made is important because it can impact on whether other steps fail, due to production rules not being applicable (see below for those conditions). We cast the problem of choosing one cell as a sorting problem and we explore three sorting criteria, according to which we select (a) (*Position* criterion) the non-terminal cell with the lowest y-coordinate in p and, in case of, tie, the one with the lowest x-coordinate; or (b) (*Recency* criterion)

Algorithm 1. Algorithm to generate a polyomino $p \in \mathcal{P}_\mathcal{G} \cup \{\varnothing\}$ from a genotype $g \in G$ using a PoCFG $\mathcal{G}$, a sorting criterion c, and an overwriting flag o.

```
 1: function DEVELOP(g, 𝒢; c, o)
 2:     p ← SINGLE(STARTINGSYMBOL(𝒢))                    ▷ init with starting symbol
 3:     s ← ∅
 4:     while true do
 5:         {(xᵢ, yᵢ)}ᵢ ← NONTERMINALCELLS(p)
 6:         if |{(xᵢ, yᵢ)}ᵢ| = 0 then                         ▷ no non terminal cells
 7:             break
 8:         end if
 9:         (x⋆, y⋆) ← FIRST(ℛ; c)                              ▷ find cell to be replaced
10:         ℛₙ ← OPTIONSFOR(p_{x⋆,y⋆}, 𝒢)  ▷ find replacing ref. pol. for p_{x⋆,y⋆} ∈ N in 𝒢
11:         (p', s) ← CHOOSE(ℛₙ, g, s)       ▷ choose one replacing referenced polyomino
12:         if p' = ∅ then                              ▷ no chosen replacing polyomino
13:             return ∅
14:         end if
15:         if ¬o ∧ ¬FITS(p', p, x⋆, y⋆) then        ▷ cannot replace p with p' at x⋆, y⋆
16:             return ∅
17:         end if
18:         p ← REPLACE(p, p', x⋆, y⋆)
19:     end while
20:     return p
21: end function
```

---

the one which has been inserted in p most recently (i.e., at the most recent iteration of the algorithm) and, in case of tie, the one selected with the Position criterion; or (c) (free *Sides* criterion) the one which has most free sides, i.e., sides on which there are no other cells, and, in case of tie, the one selected with the Position criterion. In Algorithm 1, the parameter c represents the sorting criterion determining the behavior of FIRST(). In Sect. 3.1 we compare experimentally the variants of the algorithm obtained with different criteria.

Step (iii) is the one where a production rule among the suitable ones for the target cell is chosen based on the genotype, through the CHOOSE() function. We describe three alternatives for this function in Sect. 2.3.

No-Mapping Conditions. There are two conditions according to which the development algorithm returns $\varnothing$, i.e., fails in mapping a genotype g to a polyomino $p \in \mathcal{P}_\mathcal{G}$. First, if it is not possible to choose a rule for a given replaceable cell, given a genotype g, i.e., if CHOOSE($\mathcal{R}_r, g$) returns $\varnothing$: this condition usually (see next section) corresponds to the case where g has been completely consumed. We introduced this possibility as a mechanism for avoiding endless execution of the iterative part of the development algorithm—similar mechanisms indeed exist in most of the variants of GE, such as grammatical evolution (SGE) [16] and weighted hierarchical grammatical evolution (WHGE) [2].

Second, if the chosen cell at $(x^\star, y^\star)$ of p is not replaceable using the referenced polyomino p' constituting the right-hand-side of the chosen rule, then the mapping fails. A cell $(x^\star, y^\star)$ of p is not replaceable with p' if some cells of p close to the cell are not empty and should be replaced by a corresponding cell in p'. We remark that this condition is peculiar of the case of polyominoes and does not have a counterpart in plain strings, differently from the previous one. In facts, while in the case of strings one can replace a single symbol with a sequence of symbols simply by "enlarging" the gap between the leading and trailing (with respect to the symbol being replaced) substrings, this accommodation is not possible in 2-D. In Algorithm 1, this condition is verified by FITS().

In this work, we also explore a variant of the development algorithm in which a rule is always applicable, that is, a referenced polyomino p' can always be placed at a given $(x^\star, y^\star)$, regardless of the fact that close cells are empty. In other words, in this variant we allow for *overwriting* while applying production rules—in Algorithm 1, the Boolean parameter o represents an overwriting flag.

Different Representations. We designed our development algorithm to accommodate different *representations*, i.e., different domains G for the genotype $g \in G$. Our rationale is twofold. First, we wanted to show that the algorithms is general, thanks to the decoupling of the choice of the production rule and the rest of the process. Second, we wanted to build on previous research and practice on the similar case of grammar-guided genetic programming (G3P), where different kinds of genotype (and different ways of using it) have been proposed to improve the general effectiveness of the evolutionary search, e.g., bit-strings in the early GE and WHGE, structured strings of integers in SGE.

We considered four representations; thanks to the modular nature of the algorithm, they correspond to four implementations of the CHOOSE() function of Algorithm 1. In all cases, we assume that the procedure for choosing a production rule given the genotype is stateful, that is, that subsequent invocations with the same g may give different outputs. We formalize this assumption by including a state s as argument for CHOOSE() and by making the function return a new state s, along with the chosen reference polyomino p'. We remark that the state is initialized to an empty state $\varnothing$ at every new execution of the development algorithm. The domain of the state depends on the representation.

Figure 1 shows an example of the execution of the development algorithm with three of the four representations described in detail below.

String of Integers. In this representation, a genotype g is an l-long string of integers, i.e., $G = \{1, \ldots, b\}^l \subseteq \mathbb{N}^l$, and the state s is a integer, i.e., $S = \{1, \ldots, l\} \in \mathbb{N}$, used as a counter.

Given a genotype $g = (g_1, \ldots, g_l)$, the production rules $\mathcal{R}_n = (r_1, \ldots, r_k)$ for the non-terminal n (where each r_j is a pair (n, p_j), with p_j a referenced polyomino defined over $N \cup T$), and the state s, the function CHOOSE() for this representation works as follows. Concerning the state, if s is $\varnothing$, s becomes 1, otherwise s becomes $s + 1$. Concerning the output reference polyomino p', if $s > l$, $p' = \varnothing$, otherwise $p' = p_j$, with $j = ((g_s - 1) \bmod k) + 1$.

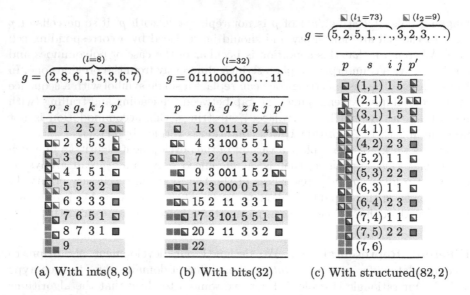

Fig. 1. Example of the development of a polyomino with the grammar of Fig. 2b and three representations (one table for each representation). Each row in the table represents one iteration of the algorithm (with the Position sorting criterion and no overwriting). The thick black border denotes in p the cell that is being replaced and in p' the reference cell.

Intuitively, here CHOOSE() consumes the genotype one integer at once and chooses the rule using the mod rule, as in the original GE.

We denote this representation with ints(l, b), b being the maximum value each genotype element may assume and l being the genotype length. That is, l and b are parameters for this parametric representation.

String of Bits. In this representation, $G = \{0, 1\}^l$ and $S = \{1, \ldots, l\} \in \mathbb{N}$.

Given $g = (g_1, \ldots, g_l)$, $\mathcal{R}_n$, and s, CHOOSE() works as follows. Concerning the state, if s is $\varnothing$, s becomes 1, otherwise s becomes $s + h$, with $h = \lceil \log_2 |\mathcal{R}_n| \rceil$. Concerning p', if $s + h > l$, $p' = \varnothing$, otherwise it (i) takes the h bits $g' = (g_s, g_{s+1}, \ldots, g_{s+h})$ in g, (ii) converts them to an integer $z \in \{1, \ldots, 2^h\}$, then (iii) returns $p' = p_j$, with $j = (z \bmod k) + 1$.

Intuitively, here CHOOSE() consumes the genotype h bits at once, with h the smallest possible to accommodate $\mathcal{R}_n$ (possibly not-consuming bits if $|\mathcal{R}_n| = 1$), and chooses the rule using the mod rule on the bit-to-integer conversion of the consumed h bits.

We denote this representation with bits(l).

String of Reals. In this representation, $G = \mathbb{R}^l$ and $S = \{1, \ldots, l\} \in \mathbb{N}$.

Given $g = (g_1, \ldots, g_l)$, $\mathcal{R}_r$, and s, CHOOSE() works as follows. Concerning the state, it works as ints(n, l). Concerning p', if $s > l$, $p' = \varnothing$, otherwise it (i) clamps down g_s in $[0, 1]$, as $h = \min(1, \max(0, g_s))$, then (ii) returns $p' = p_j$, with $j = \max(1, \lceil kg_s \rceil)$.

Intuitively, here CHOOSE() consumes the genotype one real value at once and chooses the rule based on the value clamped in $[0, 1]$ and mapped to $\{1, \ldots, |\mathcal{R}_n|\}$.

We denote this representation with reals(l).

Structured String of Integers. In this representation, a genotype is a set of strings of integers, one string for each non-terminal in the grammar, and the state is a set of counters, one for each non-terminal. Let $N = \{n_1, \ldots, n_m\}$ be the set of non-terminals and let $\mathcal{R}_{n_j} \subseteq \mathcal{R}$ be the set of production rules for the non-terminal n_j. Formally, $G = \{1, \ldots, |\mathcal{R}_{n_1}|\}^{l_1} \times \cdots \times \{1, \ldots, |\mathcal{R}_{n_m}|\}^{l_m}$ and $S = \{1, \ldots, l_1\} \times \cdots \times \{1, \ldots, l_m\}$, with $\sum_{j=1}^{j=m} l_j = l$.

We determine the number l_j of genotype elements for the j-th non-terminal of the grammar, based on the overall genotype length l, as follows. (i) We start from a bag $\mathcal{N} = \{n_1\}$ of non-terminal symbols containing the axiom only. (ii) We repeat for n_{rec} times this procedure: for each element n of $\mathcal{N}$, we consider all the rules $\mathcal{R}_n$ for n, we take all the referenced polyominoes appearing on the right-hand-side, and we add to $\mathcal{N}$ all the non-terminal symbols appearing in them (possibly multiple times). (iii) Finally, based on the content of $\mathcal{N}$, we reserve to each non-terminal n a proportion of genotype elements based on the amount of n items in $\mathcal{N}$, namely, for n_j we set $l_j = \left\lfloor l \frac{|\{n \in \mathcal{N} : n = n_j\}|}{|\mathcal{N}|} \right\rfloor$ (reasonably adjusted to have $\sum_{j=1}^{j=m} l_j = l$). The rationale of this procedure is to have a number of genotype elements suitable for performing "enough" productions with each given non-terminal, while still constraining the genotype to be l-long. The parameter n_{rec} determines how much non-terminals that are, intuitively, more recursive in the grammar take more space in the genotype.

Given $g = (g_{1,1}, \ldots, g_{1,l_1}, \ldots, g_{m,1}, \ldots, g_{m,l_m})$, the rules $\mathcal{R}_{n_i}$ for a non-terminal n_i, and $s = (s_1, \ldots, s_m)$ (or $s = \varnothing$), CHOOSE() for this representation works as follows. Concerning the state, if $s = \varnothing$, then s becomes $(1, \ldots, 1)$ (i.e., a m-long vector of ones); otherwise, if $s = (s_1, \ldots, s_m)$, then $s_i = s_i + 1$. Concerning p', if $s_i > l_i$, $p' = \varnothing$, otherwise $p' = p_j$, with $j = g_{i,s_i}$.

Intuitively, here CHOOSE() consumes the genotype one integer value at once in the portion of the genotype corresponding to the non-terminal being replaced and chooses the rule based on the "current" genotype element. Note that the mod rule here is not needed, since the domain of each genotype part exactly matches the number of rules for the corresponding non-terminal symbol.

We denote this representation with structured(l, n_{rec}).

2.4   Evolution of Polyominoes

Having defined a way to map a genotype $g \in G$ to a polyomino $p \in \mathcal{P}_G$ for a grammar $\mathcal{G}$, we can solve problems of optimization over $\mathcal{P}_G$ using evolutionary computation (EC), that is, with an EA.. Since we defined mapping variants for different kinds of genotype, we might use any EA. that is suitable for the corresponding G, e.g., evolutionary strategy (ES) [9], or maybe the more recent

OpenAI-ES [33], for the reals(l) representation and a genetic algorithm (GA) for bits(l), possibly including a linkage-exploitation mechanism [35] as done in [20]. However, for simplicity, in this work we experiment with a single EA, described below, and we leave the investigation of other EAs as future work.

Given a fitness function $f : \mathcal{P}_\mathcal{G} \to \mathbb{R}$ (we assume to tackle minimization problems, without loss of generality), we evolve polyominoes with a simple EA with two variation operators (mutation and crossover, each being representation-specific), tournament selection for parents selection, and overlapping between parents and offspring. In detail, we first initialize (with a representation-specific procedure) a population P of n_{pop} individuals. Then, we repeat n_{gen} times the following steps: (i) we generate $r_{x\text{-}over}n_{pop}$ new individuals with crossover, i.e., each one by selecting two parents from P with tournament selection (with n_{tour} size) and applying them a crossover operator; (ii) we generate $(1 - r_{x\text{-}over})n_{pop}$ new individuals with mutation, selecting the parent with tournament selection; (iii) we merge all newly generated individuals to the parents, hence obtaining a population P with $2n_{pop}$ individuals; (iv) we trim P with truncation selection, retaining the best n_{pop} individuals according to the fitness function f. At the end, we return the individual, i.e., the polyomino, with the lowest fitness.

Concerning the representation-specific initialization procedure, we simply generate each genotype g by sampling each one of its element in the proper domain with uniform probability, i.e., in $\{1, \ldots, b\}$ for ints(l, b), in $\{0, 1\}$ for bits(l), in $[0, 1]$ for reals(l), and $\{1, \ldots, |\mathcal{R}_{n_i}|\}$ (with the appropriate value for i) for structured(l, n_{rec}).

Concerning the mutation operator, we use the point-mutation, that randomly changes each genotype element to another value in the proper domain with p_{mut} probability, for ints(l, b), bits(l), and structured(l, n_{rec}). For reals(l) we use the Gaussian mutation with σ_{mut}.

Finally, concerning the crossover operator, we use the uniform crossover, that takes each element in the child genotype from one of the parents with equal probability, for all the representations.

3   Experiments and Results

We performed some experiments to: (a) compare the different variants of the development algorithm (i.e., sorting criterion and overwriting flag), (b) compare the different representations, (c) verify if our approach actually allows to evolve polyominoes towards a predefined target shape while adhering to the given grammar. Concerning the representations, we experimented with bits(l), ints($l, 4$), ints($l, 16$), reals(l), and structured($l, 2$), with different values for the genotype length l.

When comparing variants and representation, we focused on analyzing quantitatively some properties of the representation, since they allow to characterize how the search process will work [19,30]. Namely, we consider the following quantitative properties, which we measured experimentally:

Validity It measures the degree to which a genotype is mapped to a valid phenotype. Given a set G of genotypes, we obtained the corresponding bag P of phenotypes by applying our development algorithm and then we computed the validity as $\frac{1}{|G|}|\{p \in P : p \neq \varnothing\}|$.

Uniqueness It measures the degree to which different genotypes are mapped to different phenotypes. Given a set G of genotypes and the corresponding bag P of phenotypes, we computed the uniqueness as $\frac{|G|}{|P'|}$, with P' being the set of elements of P different than $\varnothing$, i.e., the valid phenotypes. Note that P may contain duplicates, while P' does not, being a set.

Locality It measures the degree to which similar genotypes are mapped to similar phenotypes. Given a sequence G of unique genotypes, the corresponding sequence P of phenotypes, and two distances d_G and d_P defined for genotypes and phenotypes, we computed the distance matrices $\boldsymbol{D}_G$ and $\boldsymbol{D}_P$ containing the distances between all pairs of elements of the two sequences and then we computed the locality as the Pearson correlation between the corresponding elements of the matrices. As d_P we used the Hamming distance of pair of polyominoes after having translated them in order to have coincident centers of mass. As d_G we used Hamming distance for bits(l), ints(l, b), and structured($l, 2$), Euclidean distance for reals(l).

For all the properties, the greater, the better.

We performed our experiments with five PoCFGs, shown in Fig. 2. They differ in the number $|\mathcal{R}|$ of rules, the number $|T|$ of terminals, and the number $|N|$ of non-terminals.

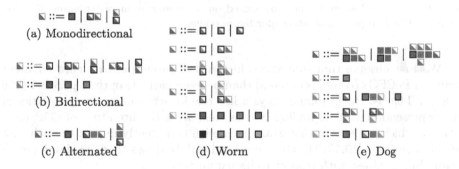

Fig. 2. The PoCFGs considered in the experiments. Half colored squares represent non-terminal symbols; fully colored squares represent terminal symbols; on the right-hand-side, a thick black border denotes the reference cell in a referenced polyomino. E.g., for the Dog grammar, $N =$, $T =$, $n_1 =$, and there are $|\mathcal{R}| = 11$ production rules. (Color figure online)

3.1   Comparison of Development Variants

For comparing the six variants of our development algorithm obtained by combining the three sorting criteria and the two values for the overwriting flags, we used the bits(l) representation, with $l \in \{10, 15, \ldots, 245, 250\}$. We performed similar experiments also for the other representations, observing qualitatively similar findings. For measuring the properties, for each l value we generated 5000 genotypes (and the corresponding phenotypes) for validity and uniqueness and 1000 genotypes for the locality. Figure 3 presents the results of this experiment.

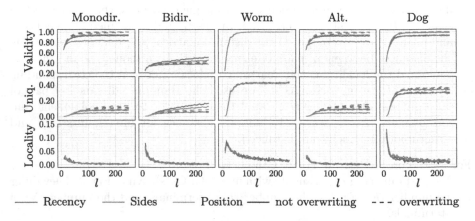

Fig. 3. Representation properties (rows of plots) for the six variants of the development algorithm (line colors and types) measured on the five grammars (columns of plots) with the bits(l) representation. (Color figure online)

We first observe that differences in terms of properties are more apparent between PoCFGs (columns of plots) than between variants of the algorithm (line colors). That is, the grammar plays a key role in determining the properties of the representation. This finding is consistent with the literature of G3P algorithms, which has shown that grammar design can greatly impact the behavior of the algorithm [10,17,24]. On the other hand, it shows that our development algorithm is robust with respect to its parameters.

By looking at the plots related to validity (first row), it is possible to see that overwriting results, in general, in a larger number of valid polyominoes. With all the grammars, except for the Bidirectional, the validity reaches its maximum for most of the combinations with overwriting and a large enough l. The reason why Bidirectional leads to lower validity, might be related to the fact that there is a higher chance of selecting a non-terminal than a terminal, when comparing to the other grammars.

Concerning uniqueness, Fig. 3 suggests that the Sides criterion tends to result in lower uniqueness and Recency in greater uniqueness. No clear and general distinctions can be made between the variants with and without overwriting.

Finally, concerning the locality, the results suggest that there are no differences among the variants. The main role is played by l: the longer the genotype, the lower the locality. This finding can be explained by the fact that long genotypes might not be used completely in the mapping process: the differences in unused parts of two genotypes would not be reflected in the corresponding phenotypes. This interplay between locality and actual usage of the genotype has already been observed in previous works and can be spotted with the help of visualization tools [23].

Based on the results of this experiment, we decided to use the Recency criterion without overwriting. Specifically, we selected the latter parameter value due to its closer alignment with the function of a grammar, namely, describing structural constraints for polyominoes.

3.2 Comparison of Representations

We compared the five representations with the same procedure of the previous experiment (and with the Recency criterion without overwriting). The results are depicted in Fig. 4.

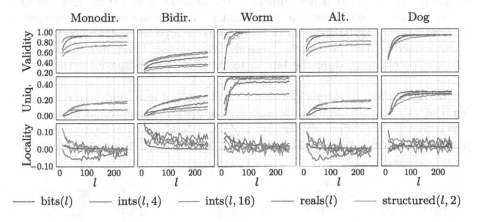

Fig. 4. Representation properties (rows of plots) for the five representations (line color) measured on the five grammars (columns of plots) with the development algorithm based on Recency without overwriting.

As in the previous analysis, the results shows that the grammar and the genotype length l impact more on the representation properties than the representation itself. However, representations differ more than development algorithm variants.

The $\mathrm{bits}(l)$ and $\mathrm{ints}(l, 4)$ representations are in general better in terms of validity. $\mathrm{structured}(l, 2)$ generates more invalid polyominoes than the other representations, likely because some portions of the genotype are not long enough—related to the n_{rec} parameter. However, larger validity does not always mean more unique phenotypes: in all the grammars except the Bidirectional and the Dog grammar, the structured$(l, 2)$ representation presents higher uniqueness.

Concerning locality, structured$(l, 2)$ and reals(l) score, in general, better. All representations present a similar trend, except bits(l), which presents a smoother line (with, however, low locality).

Although in EAs a higher locality is important, we chose to perform the subsequent experiments with the bits(l) representation, as it is the simplest one and the most similar to the original GE.

3.3   Evolution of Polyominoes

The last analysis consists in the evolution of polyominoes, as we wanted to show that the algorithm proposed can be used inside an EA to solve optimization problems. We evolved the polyominoes using the EA described in Sect. 2.4 with the following parameters: $n_{\text{pop}} = 100$, $n_{\text{gen}} = 200$, $r_{\text{x-over}} = 0.8$, $n_{\text{tour}} = 3$, $p_{\text{mut}} = 0.01$, the latter being the only representation-specific parameter. We employed the bits(500) representation with the Recency criterion and no overwriting. We used JGEA [22] for the experiments.

We built an optimization problem where the goal is to evolve a target polyomino $p^\star$. We used, as fitness function, the average of the Hamming distance of the evaluated polyomino p to the target $p^\star$ and the same distance computed without considering labels; in both cases, we translated the polyominoes in order to have their centers of mass to coincide. We employed this function, namely, also the part disregarding the labels, to facilitate the evolution of the correct shape.

We considered five target polyominoes, shown in Fig. 5, and used each of the five PoCFGs on each target polyomino. We purposely chose target polyominoes which match very differently the five PoCFGs. Note that the Dog shape is not perfectly achievable with the Dog grammar, due to the misplaced rightmost foot.

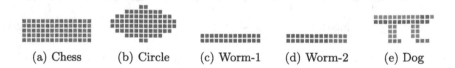

(a) Chess   (b) Circle   (c) Worm-1   (d) Worm-2   (e) Dog

Fig. 5. The five target polyominoes.

For each of the 25 combinations of grammar and target, we performed 50 evolutionary runs by varying the random seed. Figure 6 presents the results of these experiments: it shows the fitness of the best polyomino during the evolution and its size, i.e., number of cells.

By looking at the results, it is possible to see that the EA successfully identified the optimal solutions for the Worm-1 and Worm-2 by employing the Worm grammar. Similarly, the Dog grammar greatly outperformed the other grammars in solving the Dog problem. This observation highlights the significance of

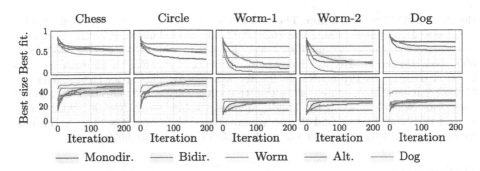

Fig. 6. Fitness and size (row of plots) of the best polyomino during the evolution for the five problems (column of plots) using the five grammars (color line) with the development algorithm based on Recency without overwriting and the bits(500) representation. The shaded area corresponds to the interquartile range, the line to the median across 50 runs. (Color figure online)

a well-designed grammar, not only to specify structural constraints, but also to incorporate domain-specific knowledge about the problem.

When employing a grammar not specifically designed for the problem at hand, such as the Alternated grammar for Circle, Worm-1, Worm-2, and Dog problems, the EA seems to become trapped in local minima after a few iterations. This is indicated by the size of the fittest individual, which remains unchanged, highlighting the EA difficulty to explore the solution space effectively under these circumstances.

All to all, these experiments show that it is possible to evolve a polyomino towards a specific target while keeping it satisfying some predefined constraints, encoded in a user defined PoCFG.

4   Conclusions

This work introduces the concept of PoCFGs and a novel approach to generate polyominoes that meet predetermined requirements, defined by a PoCFG. The experimental results align with existing literature of G3P algorithms, highlighting the importance of grammar design [10,17,24], as differences in terms of representation properties are more apparent between PoCFG, than between variants of the algorithm. Additionally, we show the adaptability and potential of our approach by integrating the algorithm within EA to evolve polyominoes towards a specific target while satisfying some predefined constraints encoded in a designed PoCFG.

Future work aims to explore the applicability of the algorithm by evolving polyominoes in more complex problems, such as the generation and evolution of modular robots [28], maps for games [11], or DNA shapes [26].

Acknowledgements. This research is the result of the collaboration with the Department of Engineering and Architecture of the University of Trieste, Italy; supported by the 2023 SPECIES scholarship. The first author is funded by FCT - Foundation for Science and Technology, under the grant 2022.10174.BD. This work was supported by the Portuguese Recovery and Resilience Plan (PRR) through project C645008882-00000055, Center for Responsible AI, by the FCT, I.P./MCTES through national funds (PIDDAC), by Project No. 7059 - Neuraspace - AI fights Space Debris, reference C644877546-00000020, supported by the RRP - Recovery and Resilience Plan and the European Next Generation EU Funds, following Notice No. 02/C05-i01/2022, Component 5 - Capitalization and Business Innovation - Mobilizing Agendas for Business Innovation, and within the scope of CISUC R&D Unit - UIDB/00326/2020.

References

1. Barequet, G., Golomb, S.W., Klarner, D.A.: Polyominoes. In: Handbook of Discrete and Computational Geometry, pp. 359–380, Chapman and Hall/CRC (2017)
2. Bartoli, A., Castelli, M., Medvet, E.: Weighted hierarchical grammatical evolution. IEEE Trans. Cybern. **50**(2), 476–488 (2018)
3. Conway, A.: Enumerating 2D percolation series by the finite-lattice method: theory. J. Phys. A: Math. Gen. **28**(2), 335 (1995)
4. Fernau, H., Schmid, M.L., Subramanian, K.G.: Two-dimensional pattern languages. In: Workshop on Non-classical Models for Automata and Applications (2017)
5. Fukuda, H., Kanomata, C., Mutoh, N., Nakamura, G., Schattschneider, D.: Polyominoes and polyiamonds as fundamental domains of isohedral tilings with rotational symmetry. Symmetry **3**(4), 828–851 (2011). ISSN 2073-8994
6. Giammarresi, D., Restivo, A.: Two-dimensional languages. In: Rozenberg, G., Salomaa, A. (eds.) Handbook of Formal Languages, pp. 215–267. Springer, Heidelberg (1997). https://doi.org/10.1007/978-3-642-59126-6_4. ISBN 978-3-642-59126-6
7. Golomb, S.W., Klarner, D.A.: Polyominoes. In: Goodman, J.E., O'Rourke, J. (eds.) Handbook of Discrete and Computational Geometry, 2nd edn., pp. 331–352. Chapman and Hall/CRC (2004)
8. Grimmett, G.: What is Percolation? In: Grimmett, G. (ed.) Percolation. Grundlehren der mathematischen Wissenschaften, vol. 321, pp. 1–31. Springer, Heidelberg (1999). https://doi.org/10.1007/978-3-662-03981-6_1. ISBN 978-3-662-03981-6
9. Hansen, N., Arnold, D.V., Auger, A.: Evolution strategies. In: Kacprzyk, J., Pedrycz, W. (eds.) Springer Handbook of Computational Intelligence, pp. 871–898. Springer, Heidelberg (2015). https://doi.org/10.1007/978-3-662-43505-2_44
10. Harper, R.: GE, explosive grammars and the lasting legacy of bad initialisation. In: IEEE Congress on Evolutionary Computation. IEEE (2010)
11. Johnson, L., Yannakakis, G.N., Togelius, J.: Cellular automata for real-time generation of infinite cave levels. In: Proceedings of the 2010 Workshop on Procedural Content Generation in Games, pp. 1–4 (2010)
12. Knight, T., Stiny, G.: Making grammars: from computing with shapes to computing with things. Des. Stud. **41**, 8–28 (2015)
13. Knuth, D.E.: Dancing links. arXiv preprint cs/0011047 (2000)
14. Křivka, Z., Martín-Vide, C., Meduna, A., Subramanian, K.G.: A variant of pure two-dimensional context-free grammars generating picture languages. In: Barneva, R.P., Brimkov, V.E., Šlapal, J. (eds.) IWCIA 2014. LNCS, vol. 8466, pp. 123–133. Springer, Cham (2014). https://doi.org/10.1007/978-3-319-07148-0_12. ISBN 978-3-319-07148-0

15. Lavirotte, S., Pottier, L.: Optical formula recognition. In: Proceedings of the Fourth International Conference on Document Analysis and Recognition, vol. 1, pp. 357–361. IEEE (1997)

16. Lourenço, N., Pereira, F.B., Costa, E.: Unveiling the properties of structured grammatical evolution. Genet. Program Evolvable Mach. **17**, 251–289 (2016)

17. Manzoni, L., Bartoli, A., Castelli, M., Gonçalves, I., Medvet, E.: Specializing context-free grammars with a (1+ 1)-EA. IEEE Trans. Evol. Comput. **24**(5), 960–973 (2020)

18. Matz, O.: Regular expressions and context-free grammars for picture languages. In: Reischuk, R., Morvan, M. (eds.) STACS 1997. LNCS, vol. 1200, pp. 283–294. Springer, Heidelberg (1997). https://doi.org/10.1007/BFb0023466. ISBN 978-3-540-68342-1

19. Medvet, E.: A comparative analysis of dynamic locality and redundancy in grammatical evolution. In: McDermott, J., Castelli, M., Sekanina, L., Haasdijk, E., García-Sánchez, P. (eds.) EuroGP 2017. LNCS, vol. 10196, pp. 326–342. Springer, Cham (2017). https://doi.org/10.1007/978-3-319-55696-3_21

20. Medvet, E., Bartoli, A., De Lorenzo, A., Tarlao, F.: GOMGE: gene-pool optimal mixing on grammatical evolution. In: Auger, A., Fonseca, C.M., Lourenço, N., Machado, P., Paquete, L., Whitley, D. (eds.) PPSN 2018, Part I. LNCS, vol. 11101, pp. 223–235. Springer, Cham (2018). https://doi.org/10.1007/978-3-319-99253-2_18

21. Medvet, E., Nadizar, G.: GP for continuous control: teacher or learner? The case of simulated modular soft robots. In: Winkler, S., Trujillo, L., Ofria, C., Hu, T. (eds.) Genetic Programming Theory and Practice XX. Genetic and Evolutionary Computation, pp. 203–224. Springer, Singapore (2023). https://doi.org/10.1007/978-981-99-8413-8_11

22. Medvet, E., Nadizar, G., Manzoni, L.: JGEA: a modular java framework for experimenting with evolutionary computation. In: Proceedings of the Genetic and Evolutionary Computation Conference Companion, pp. 2009–2018 (2022)

23. Medvet, E., Virgolin, M., Castelli, M., Bosman, P.A., Gonçalves, I., Tušar, T.: Unveiling evolutionary algorithm representation with du maps. Genet. Program Evolvable Mach. **19**, 351–389 (2018)

24. Nicolau, M., Agapitos, A.: Understanding grammatical evolution: grammar design. In: Ryan, C., O'Neill, M., Collins, J.J. (eds.) Handbook of Grammatical Evolution, pp. 23–53. Springer, Cham (2018). https://doi.org/10.1007/978-3-319-78717-6_2

25. Noya, E., Benedí, J.M., Sánchez, J.A., Anitei, D.: Discriminative learning of two-dimensional probabilistic context-free grammars for mathematical expression recognition and retrieval. In: Pinho, A.J., Georgieva, P., Teixeira, L.F., Sánchez, J.A. (eds.) IbPRIA 2022. LNCS, vol. 13256, pp. 333–347. Springer, Cham (2022). https://doi.org/10.1007/978-3-031-04881-4_27

26. Ong, H.S., Syafiq-Rahim, M., Kasim, N.H.A., Firdaus-Raih, M., Ramlan, E.I.: Self-assembly programming of DNA polyominoes. J. Biotechnol. **236**, 141–151 (2016). ISSN 0168-1656

27. Ota, P.A.: Mosaic grammars. Pattern Recogn. **7**(1–2), 61–65 (1975)

28. Pigozzi, F., Medvet, E., Bartoli, A., Rochelli, M.: Factors impacting diversity and effectiveness of evolved modular robots. ACM Trans. Evol. Learn. **3**(1), 1–33 (2023)

29. Prusa, D., Hlavá, V.: 2D context-free grammars: Mathematical formulae recognition. In: Prague Stringology Conference (2006)

30. Rothlauf, F., Goldberg, D.E.: Redundant representations in evolutionary computation. Evol. Comput. **11**(4), 381–415 (2003)

31. Ryan, C., Collins, J.J., Neill, M.O.: Grammatical evolution: evolving programs for an arbitrary language. In: Banzhaf, W., Poli, R., Schoenauer, M., Fogarty, T.C. (eds.) EuroGP 1998. LNCS, vol. 1391, pp. 83–96. Springer, Heidelberg (1998). https://doi.org/10.1007/BFb0055930
32. Sakai, I.: Syntax in universal translation. In: Proceedings of the International Conference on Machine Translation and Applied Language Analysis (1961)
33. Salimans, T., Ho, J., Chen, X., Sidor, S., Sutskever, I.: Evolution strategies as a scalable alternative to reinforcement learning. arXiv preprint arXiv:1703.03864 (2017)
34. Subramanian, K., Ali, R.M., Geethalakshmi, M., Nagar, A.K.: Pure 2D picture grammars and languages. Discret. Appl. Math. **157**(16), 3401–3411 (2009)
35. Thierens, D., Bosman, P.A.: Optimal mixing evolutionary algorithms. In: Proceedings of the 13th Annual Conference on Genetic and Evolutionary Computation, pp. 617–624 (2011)
36. Whittington, S.G., Soteros, C.E.: Lattice animals: rigorous results and wild guesses (1990)
37. Winslow, A.: Staged self-assembly and polyomino context-free grammars. Nat. Comput. **14**(2), 293–302 (2015)

Naturally Interpretable Control Policies via Graph-Based Genetic Programming

Giorgia Nadizar[1] , Eric Medvet[2](✉) , and Dennis G. Wilson[3]

[1] Department of Mathematics and Geosciences, University of Trieste, Trieste, Italy
[2] Department of Engineering and Architecture, University of Trieste, Trieste, Italy
emedvet@units.it
[3] ISAE-SUPAERO, University of Toulouse, Toulouse, France

Abstract. In most high-risk applications, interpretability is crucial for ensuring system safety and trust. However, existing research often relies on hard-to-understand, highly parameterized models, such as neural networks. In this paper, we focus on the problem of policy search in continuous observations and actions spaces. We leverage two graph-based Genetic Programming (GP) techniques—Cartesian Genetic Programming (CGP) and Linear Genetic Programming (LGP)—to develop effective yet interpretable control policies. Our experimental evaluation on eight continuous robotic control benchmarks shows competitive results compared to state-of-the-art Reinforcement Learning (RL) algorithms. Moreover, we find that graph-based GP tends towards small, interpretable graphs even when competitive with RL. By examining these graphs, we are able to explain the discovered policies, paving the way for trustworthy AI in the domain of continuous control.

Keywords: Graph-based Genetic Programming · Cartesian Genetic Programming · Linear Genetic Programming · Interpretable Policy · Continuous Control

1 Introduction

In the last decade, Artificial Neural Networks (ANNs) have achieved outstanding results in a myriad of domains, surpassing human performance even in scenarios where fast decision making (e.g., drone control [20]) or long-term planning (e.g., strategy games [41]) are needed. This success comes in conjunction with the multiplication of available optimization techniques for ANNs [40], which have become sample-efficient, oftentimes robust, and even transferable to the real world [38].

However, ANNs carry along a natural downside: their large structures, with as many as billions of parameters, remain often completely unintelligible for human observers. In fact, although several post-hoc explanation methods have been recently put forward [1,15,35], ANN-based controllers remain inherently black-box models, the true functioning of which can only be surmised through statistical inference.

ⓒ The Author(s), under exclusive license to Springer Nature Switzerland AG 2024
M. Giacobini et al. (Eds.): EuroGP 2024, LNCS 14631, pp. 73–89, 2024.
https://doi.org/10.1007/978-3-031-56957-9_5

In opposition to this paradigm, there has been renewed interest in the development of *interpretable models*, i.e., models which can be "directly" understood, without the need of additional proxies or post-hoc procedures. Even in the absence of a shared definition of interpretability [28,34,46], models which comply to some notion of interpretability are considered preferable w.r.t. black-boxes [37], as they allow humans to directly conceive their working rationales. Not surprisingly, interpretability plays a fundamental role principally in high-stakes applications, where transparency is among the key factors for trustworthiness [14].

In particular, the domain of robotics is starting to witness a call for more transparent systems, as most people are reluctant to the adoption of autonomous robots in the wild, especially when their control algorithms remain obscure. More generally, this need affects the broader family of control, where the objective is to find a control law that links inputs and outputs towards the achievement of some goal.

In this work, we aim at discovering efficient yet interpretable control laws, by leveraging functional graphs, which are well-suited for capturing complex mappings between observations and actions while maintaining a fair degree of interpretability when their size is contained. For searching the space of graphs, we rely on two graph-based Genetic Programming (GP) techniques, Cartesian Genetic Programming (CGP) and Linear Genetic Programming (LGP), targeting our optimization at performance, and letting interpretability naturally emerge.

Our experiments, conducted on eight continuous control problems from the Mujoco suite [42], reveal that the obtained policies are often comparable with the state-of-the-art in terms of performance, while being naturally more interpretable, even in the absence of an explicit incentive for interpretability. Moreover, the results achieved via graph-based GP appear more consistent and less sensitive to hyper-parameters tuning than those obtained with Reinforcement Learning (RL).

In the remainder of the paper, we briefly survey some relevant related works in Sect. 2, before describing our methodology and the needed background notions in Sect. 3. Last, we detail our experiments and discuss their results in Sects. 4 and 5, before drawing conclusions in Sect. 6.

2   Related Works

In recent years, the field of Explainable Reinforcement Learning (XRL) has gained notable traction [35,47], mostly abiding to the necessity of understanding the rationale underlying decisions made by artificial agents intended for deployment in the wild. Taking a step even further, considerable effort has been put into developing policies that are understandable per se, paving the way towards interpretable RL [13]. In this context, several proposals for interpretable policies have appeared, ranging from symbolic policies to decision trees, together with a wide range of methods for obtaining them [27]. Among them, a popular choice consists in leveraging evolutionary computation techniques in combination with classical RL approaches [50]. In particular, GP and its derivatives are naturally suited for yielding interpretable policies: from its very origins in the '90 s, Koza and Rice [26] proved how GP could be used for searching a policy to control a robot.

Following this seminal work, GP has often been used to solve various control tasks, be it via direct policy search or through imitation learning, where an agent learns by mimicking the actions/behavior of an expert, i.e., of a well-performing policy. For instance, Verma et al. [43] distilled an ANN policy into an interpretable program for The Open Racing Car Simulator (TORCS), obtaining interpretability and robustness at the cost of a slight performance degradation. However, although imitation learning partially addresses the poor sample efficiency of GP [40], it is not always as effective as direct policy search. In particular, Hein et al. [17] have considered a model-based RL setting and have shown that symbolic policies learned via direct interaction with the world model outperform those obtained with symbolic regression to imitate the behavior of a pre-trained ANN. Furthermore, sometimes imitation learning can even yield completely unprofitable policies, especially when the robot/environment interactions are too complex to be captured via imitation only [31].

Instead, many works have used GP for direct policy search, mostly focusing on scenarios with discrete action spaces. Custode and Iacca [6] and Ferigo et al. [9,10] have evolved decision trees, using GP hybridized with RL and quality-diversity optimization, respectively, achieving interpretable and effective policies on simple control tasks, such as cart pole and mountain car. Video games have often been used as test-beds, given the difficulty of processing high-dimensional visual inputs as the full gamut of pixels on a screen. In this context, most of the efforts have been devoted to solving the Atari benchmark [30], e.g., with co-evolved decision trees [7], or with Tangled Program Graphss (TPGs) [21], even obtaining a single program able to solve multiple games [22,24]. Closer to our work, Wilson et al. [48] relied on mixed-type CGP for finding simple and well-performing policies for playing the Atari games. Interestingly, as in our case, the simplicity of the policies was not promoted explicitly during evolution, but emerged naturally thanks to the representation employed.

Moving even closer to our study, fewer works have succeeded in solving problems in continuous and multi-dimensional action spaces, as these problems require finding and exploiting interdependencies between outputs to achieve coordination. For instance, Medvet and Nadizar [31] have experimented with an ensemble of GP trees for controlling the actuation values of a simulated soft robot, even matching the performance of ANNs. CGP has also been applied to continuous control problems: Wilson et al. [49] proposed a positional variant of CGP where the position of nodes was subject to evolution, and tested it on 9 benchmark problems, including three Mujoco environments. This study is again strongly linked with ours, although we consider the standard CGP variant on more environments, and obtain generally better results. Videau et al. [44] have also employed multi-tree GP for solving a set of Mujoco environments and have also used LGP to allow information sharing among outputs. This work is closely related to ours, although they consider a bi-objective optimization where they explicitly reward policy simplicity, while we let program simplicity emerge from evolution only. Moreover, we consider a larger set of environments. More recently, LGP has also been successfully used on a complex robotic agent, the Laikago robot, for discovering robust and resilient policies which are also easy to understand, being expressed as transparent computer programs [23]. Last, Amaral et al. [2] have shown the potential of Symbiotic Bid-based GP hybridized with TPGs for a continuous control locomotion task.

GP has been shown as a useful means of discovering interpretable policies on a number of environments. In this work, we hope to further this literature by demonstrating that graph-based GP can naturally find simple, interpretable policies for continuous control on complex simulated robotics tasks.

3   Graph-Based Genetic Programming for Continuous Control

We concentrate on the domain of continuous control, with the aim of finding an *interpretable* yet effective controller, whose actions will steer the system towards the achievement of a certain goal. Namely, we consider those tasks where both the observations and the possible actions are real-valued and multivariate, in a discrete-time simulated environment.

More formally, an environment is a Markov Decision Process (MDP) [36], i.e., a tuple $(\mathcal{S}, \mathcal{A}, p, r)$, where $\mathcal{S} \subseteq \mathbb{R}^n$ and $\mathcal{A} \subseteq \mathbb{R}^m$ are the state and action spaces, respectively, $p : \mathcal{S} \times \mathcal{A} \to \mathcal{S}$ is the function describing the dynamics of the environment, and $r : \mathcal{S} \times \mathcal{A} \to \mathbb{R}$ is the reward function. We consider partial observability, thus we introduce an observation space $\mathcal{O} \subseteq \mathbb{R}^q$, with $q \leq n$, and an observation function $\phi : \mathcal{S} \to \mathcal{O}$ mapping states to the corresponding observations. At each simulation time step t, the environment is in a state $\boldsymbol{s}_t \in \mathcal{S}$, the agent observes a subset of the current state, $\phi(\boldsymbol{s}_t) = \boldsymbol{o}_t \in \mathcal{O}$, and takes an action $\boldsymbol{a}_t \in \mathcal{A}$ according to a *policy* $\pi : \mathcal{O} \to \mathcal{A}$, which results in the system changing into state $\boldsymbol{s}_{t+1}$ with probability $p(\boldsymbol{s}_{t+1}|\boldsymbol{s}_t, \boldsymbol{a}_t)$ and the agent receiving a reward $r(\boldsymbol{s}_t, \boldsymbol{a}_t)$. The objective consists in finding a policy which maximizes the cumulative reward obtainable in a simulated episode of duration T, $R_T(\pi) = \sum_{t=0}^{T} r(\boldsymbol{s}_t, \pi(\boldsymbol{o}_t))$, starting from an initial state $\boldsymbol{s}_0$. While we consider the formalization of an MDP, in this study we use optimization methods based only on the total episode reward $R_T(\pi)$. Unifying the terminology with the previous paragraph, π is the controller we search for, which determines the actions in the environment.

In our case, π is a graph which represents a multi-variate function. This family of policies can be inherently interpretable because each graph results from the high-level composition of simple functions, e.g., $+$, $-$, or sin, and is constrained in size. We rely on two flavours of graph-based GP for searching the space of graphs, namely CGP and LGP. We define the fitness of a policy graph π as the aforementioned cumulative reward, $f(\pi) := R_T(\pi)$, which evolution maximizes.

3.1 Graph-Based Genetic Programming

Although GP was primarily born as a technique for evolving computer programs, internally represented as trees [25], throughout the years graph-based alternatives have gained notable traction [11]. Here we consider two variants of graph-based GP, CGP [33] and LGP [19], which share a linear integer genotype and a phenotype that can be interpreted as a Directed Acyclic Graph (DAG). We describe them in further details in Sects. 3.2 and 3.3, while we devote the remainder of this section to describing the Genetic Algorithm (GA) we leverage for both of them.

The first step of the optimization process consists in the initialization of a population of size n_{pop}, i.e., in the generation of n_{pop} genotypes; each genotype θ is then mapped into a policy graph π_θ, and evaluated according to the fitness function $f(\pi_\theta)$. Then, the GA proceeds by iterating the following steps for n_{gen} generations. First, n_p parents are selected from the population with tournament selection (with size n_{tour}). Then, the collection of offspring, also of size n_p, is generated by applying some genetic operators. Next, the offspring is encoded and evaluated. Last, the offspring is merged with the best $n_{pop} - n_p$ individuals of the parent population (selected with truncation selection), obtaining the new population, which is used to start the following loop.

The genotype, the encoding procedure, and the genetic operators employed differ between CGP and LGP, as we describe in the following.

3.2 Cartesian Genetic Programming

CGP represents programs as grids of nodes in a Cartesian plane [32,33]. Each node represents a function, which can use either the outputs of the nodes of the previous layers or the program inputs as arguments. The final outputs of the program are collected from the outputs of n_{out} selected nodes.

Here, without loss of generality, we consider a uni-dimensional grid, i.e., a sequence of nodes of length n_{nodes}. Hence, to fully determine a CGP graph, we need (1) n_{out} indexes to specify the program outputs, and (2) n_{nodes} tuples $(h, i_1, \ldots, i_{m_{ar}})$, specifying the function and its arguments for each node, where m_{ar} is the maximum arity of the functions available in the function set H, 2 in our case. We remark that for each of the tuples, h is an index bounded by the cardinality of H, whereas each input index can range from 0 to $n_{in} + j - 1$, where j indicates the current node position. Thus, the genome of a CGP graph is a sequence of bounded integers of size $n_{out} + (1 + m_{ar})n_{nodes}$.

To initialize each genome, we sample a uniform distribution over the allowed integer values for each position. As a genetic operator, we use int-flip mutation, i.e., we sample a new integer value in the allowed interval, with a different probability for node inputs p_i, node functions p_f, and program outputs p_o. While CGP traditionally used a $(1 + \lambda)$ EA for evolution [33], the use of a GA with multiple elites, as we have done here, has shown similar results [32] and allows for greater parallelization and a direct comparison with LGP.

3.3   Linear Genetic Programming

In LGP, programs are lists of instructions from a programming language, where the result of each instruction is assigned to a register from a predefined set [4]. The inputs of the program are copied into the first n_{in} registers, whereas the outputs are taken from the last n_{out} registers. If we consider the information flow in a LGP program, we can interpret it as a DAG, and hence LGP falls under the category of graph-based GP [11].

To describe a LGP program, each instruction is determined by specifying (1) the index of the register to be assigned, (2) the function to execute from those available in H, and (3) the sequence of arguments to be passed to the function $(i_1, \ldots, i_{m_{ar}})$, expressed in terms of register indexes ($m_{ar} = 2$ being the maximum arity of functions in H). Hence, to fully determine a program of n_{lines} lines, the genome is a sequence of bounded integers of size $(2 + m_{ar})n_{lines}$. In this case the bounds for register indexes are constituted by the amount of available registers n_{reg}, as even those not explicitly assigned before are initialized to 0 and are available to select.

To generate each genome, we sample a uniform distribution over the allowed integer values for each position. Regarding the genetic operators, we apply one-point crossover (constraining instructions to be atomic) followed by int-flip mutation, with a different probability for the left-hand side of each program line, i.e., the registers to be assigned, p_a, functions p_f, and function inputs p_i. We provide hyperparameters and other experimental details below.

4   Experimental Evaluation

In our experimental evaluation we address the following research question: *"can graph-based GP yield effective yet interpretable policies for continuous control tasks?"* To this end, we perform several optimizations for both CGP and LGP on 8 continuous control tasks from the Mujoco suite. To have meaningful effectiveness baselines, we compare the performance of the resulting graph policies with those obtained with two state-of-the-art RL algorithms which represent policies using ANNs. Moreover, we measure the complexity of the graphs obtained by GP, and analyze example policies in detail to appraise their interpretability.

We open-source our code and our experimental results at https://github.com/giorgia-nadizar/cgpax.

4.1   Continuous Control Benchmark Tasks

We employ 8 benchmark continuous control tasks from the Mujoco suite [42]. Namely, we consider environments of growing complexity in terms of input and output space dimensionality, as we summarize in Table 1.

Specifically, we consider two balancing tasks, *inverted pendulum* and *inverted double pendulum*, where the controller is rewarded for preventing the pendulum from falling (and also penalized for excessive movements for the inverted double

Table 1. Observation and action space sizes for the considered problems.

Environment	Obs. $\mathcal{O}$	Act. $\mathcal{A}$	Environment	Obs. $\mathcal{O}$	Act. $\mathcal{A}$
Inverted pendulum	$\mathbb{R}^4$	$[-3, 3]$	Hopper	$\mathbb{R}^{11}$	$[-1, 1]^3$
Inverted double pendulum	$\mathbb{R}^8$	$[-1, 1]$	Walker2d	$\mathbb{R}^{17}$	$[-1, 1]^6$
Reacher	$\mathbb{R}^{11}$	$[-1, 1]^2$	Half cheetah	$\mathbb{R}^{18}$	$[-1, 1]^6$
Swimmer	$\mathbb{R}^8$	$[-1, 1]^2$	Ant	$\mathbb{R}^{27}$	$[-1, 1]^8$

pendulum). We also incorporate a target-aiming task, *reacher*, where a robotic arm has to reach an object and is rewarded for getting as close to the object as possible and penalized for excessive movements. Last, we include 5 locomotion tasks, *swimmer*, *hopper*, *walker2d*, *half cheetah*, and *ant*, which all use a positive reward for the distance covered and a penalty term for excessive movements. Among the locomotion problems, hopper, walker2d, and ant can be unstable; they therefore have an additional reward term for maintaining their balance throughout the simulation. For these problems, if the agent falls to the ground, the episode is terminated earlier.

For all tasks, we use the Brax [12] implementation which leverages parallelism on GPU accelerators with JAX [3]. We use the v1 simulation engine of Brax and set all parameters to their default values. We run all episodes to a duration $T = 1000$ time steps.

4.2 Reinforcement Learning Baselines

As baselines for comparing the performance of the graph policies, we use two state-of-the-art RL algorithms for the optimization of ANNs, Proximal Policy Optimization (PPO) and Soft Actor Critic (SAC).

PPO [39] is an on-policy RL algorithm that optimizes agent policies within a "trust region", balancing the exploration-exploitation trade-off. It achieves this by optimizing a first-order approximation of the expected reward while ensuring that policy updates remain close to the actual values.

SAC [16] is an off-policy RL algorithm that balances the maximization of two objectives: the expected cumulative reward and the policy entropy. By simultaneously promoting exploration through entropy maximization and exploitation through return maximization, SAC ensures robust learning in complex environments with continuous action spaces.

These two algorithms are considered state-of-the-art for deep RL and are the default algorithms for the Brax library. We use the Brax library implementations for both PPO and SAC.

4.3 Parameter Settings

For the GP evolutionary loop we set $n_{\text{pop}} = 100$, $n_p = 90$, and $n_{\text{tour}} = 3$. Concerning the amount of iterations performed n_{gen}, instead, we have a different value for each problem, depending on how difficult the optimization task is. Namely, for inverted pendulum we set $n_{\text{gen}} = 10$, for half-cheetah we set $n_{\text{gen}} = 500$, for inverted double pendulum, reacher, and swimmer we set $n_{\text{gen}} = 1000$, for hopper and walker2d we set $n_{\text{gen}} = 1500$, and for ant we set $n_{\text{gen}} = 2500$. Regarding the representation specific parameters, we set $n_{\text{nodes}} = 50$, $p_i = p_f = 0.1$, and $p_o = 0.3$ for CGP, while we set $n_{\text{reg}} = n_{\text{in}} + n_{\text{out}} + 5$, $n_{\text{lines}} = 15$, $p_a = 0.3$, and $p_f = p_i = 0.1$ for LGP.

For both CGP and CGP we consider the following function set $H = \{\bullet + \bullet, \bullet - \bullet, \bullet \times \bullet, \bullet \div^* \bullet, |\bullet|, \exp \bullet, \sin \bullet, \cos \bullet, \log^* \bullet, \bullet < \bullet, \bullet > \bullet\}$, where $\bullet$ represents an operand, and operators marked with $*$ are protected. The last two functions are Boolean functions, where the output is 1 if the condition is satisfied and 0 otherwise. Moreover, we add two constant inputs to each observation, namely $\{0.1, 1\}$, thus increasing all observation spaces dimensionalities displayed in Table 1 by 2.

For the RL algorithms, we optimize an ANN with 4 layers of size 32 with SiLU activation for PPO, and an ANN of 2 layers of size 256 with ReLU activation for SAC, following the default provided in Brax. Concerning the algorithm specific hyper-parameters we apply the default ones from the Brax implementations of both PPO and SAC with a few minor variations (we report the full parameter list at https://github.com/giorgia-nadizar/cgpax/blob/main/rl_run.py).

In all cases, we repeat the optimization for 10 independent times to ensure results consistency. Furthermore, for the GP techniques we evaluate each individual on 5 distinct episodes and consider the median reward across these episodes as fitness, in order to make the evolutionary search more stable. For the same reason, we also re-evaluate the elite individuals upon merging them into the offspring population.

Last, when analyzing results in a comparative manner we perform a Mann-Whitney U test between pairs of distributions, considering equivalence between the two as null hypothesis. When performing multiple pairwise comparisons, we apply the Bonferroni correction, thus dividing the significance level $\alpha = 0.05$ by the number of pairwise comparisons computed.

5 Experimental Results and Discussion

5.1 Performance Results

We report the results in terms of performance in Fig. 1. We show the distribution across 10 runs of the cumulative reward $R_T(\pi^*)$ obtained by the best policy π^* discovered at the end of the optimization on episodes of $T = 1000$ time steps. We divide our data by environment and by optimization technique to ease the visual comparison of results. We also show the p-values resulting from the pairwise Mann-Whitney U tests performed in Table 2, again dividing data by environment.

The first observation we can make regards the comparison between the two employed graph-based GP techniques: in all the considered environments they achieve similar results, with no statistical difference ever observed among the

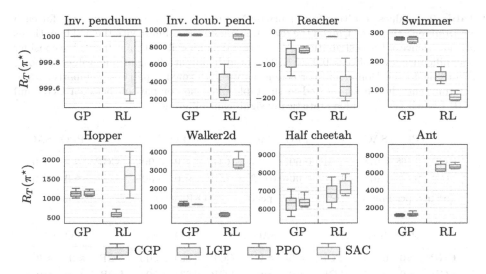

Fig. 1. Box plots of the cumulative reward $R_T(\pi^*)$ collected by the best policy discovered at the end of optimization π^* in an episode of duration $T = 1000$ time steps.

pairs of distributions. We also note that the results are usually consistent for both CGP and LGP, as opposed to the two glsrl techniques, where the distributions appear generally more spread and with greater differences between the two considered algorithms, PPO and SAC. This suggests that graph GP is more consistent than RL, which is a desirable property for achieving robust results. We speculate this partially descends from the high sensitivity of these RL algorithms to the chosen hyper-parameters.

For a more detailed comparative analysis, we return to the first part of our research question, concerning the effectiveness of the graph policies found. In the scope of this work, we consider a policy to be effective if it is not significantly worse than the examined RL baselines. Given the reduction in the number of parameters used and the subsequent interpretability gain, we consider policies which achieve similar, but not necessarily better, results to state-of-the-art policies as effective. Among the studied environments, graph policies are not significantly worse than either RL policy in 50 % of cases (inverted pendulum, inverted double pendulum, swimmer, and hopper), while they are not worse than at least one of the RL ones in three more cases (reacher, walker2d, and half cheetah). In fact, the only environment where graph policies are not able to match neither of the RL algorithms is the ant, likely because of the higher dimensionality of both action and observation spaces, and of the known difficulty of graph GP of finding good mappings between high-dimensional spaces. Thus, we can give a generally affirmative answer regarding the effectiveness of both CGP and LGP policies.

Interestingly, in some cases, graph GP is even able to significantly outperform RL. Namely, in four environments (inverted double pendulum, reacher, hopper, and walker2d) both CGP and LGP significantly surpass one of the two RL

Table 2. p-values resulting from pairwise Mann-Whitney U tests comparing, for each environment, the algorithm on the row with that on the column with the null hypothesis of equivalence of the distributions. We write ≈ 0 for all cells with $p < 0.001$. We consider a significance level of $\alpha = 0.05/5 = 0.01$ (and highlight all significant values in bold) according to the Bonferroni correction, as we performed 5 pairwise comparisons per environment: the ones reported in the table plus the comparison between CGP and LGP (which never yielded statistically significant differences).

	PPO	SAC		PPO	SAC		PPO	SAC		PPO	SAC
CGP	0.168	0.078	CGP	**0.005**	0.571	CGP	$\approx$**0**	**0.003**	CGP	$\approx$**0**	$\approx$**0**
LGP	0.168	0.078	LGP	**0.006**	0.623	LGP	$\approx$**0**	**0.003**	LGP	$\approx$**0**	$\approx$**0**
(a) Inv. pendulum			(b) Inv. doub. pend.			(c) Reacher			(d) Swimmer		

	PPO	SAC		PPO	SAC		PPO	SAC		PPO	SAC
CGP	$\approx$**0**	0.021	CGP	$\approx$**0**	$\approx$**0**	CGP	0.064	**0.003**	CGP	$\approx$**0**	$\approx$**0**
LGP	$\approx$**0**	0.017	LGP	$\approx$**0**	$\approx$**0**	LGP	0.031	**0.001**	LGP	$\approx$**0**	$\approx$**0**
(e) Hopper			(f) Walker2d			(g) Half cheetah			(h) Ant		

algorithms, whereas on swimmer both GP methods neatly outperform both RL algorithms. Arguably, some of these outcomes might come from sub-optimal parameter tuning for the poor-performing RL algorithm, which could potentially yield better results with more tweaking. However, this further corroborates our previous point concerning the robustness of GP results w.r.t. hyper-parameter tuning, unlike RL which is highly sensitive to changes in hyperparameters [18].

To gain deeper insight on the evolutionary process underlying the GP policy search, we also display the progression of the fitness of the best individual in the population, i.e., the cumulative reward $R_T(\pi^\star)$, at each iteration in Fig. 2. For all the reported plots, we note that evolution has achieved a plateau, where additional generations would likely result in null or negligible fitness improvements. Moreover, in most cases we can distinguish an initial phase where the fitness grows steeply, followed by a longer phase with minor improvements. This highlights how evolution tends to converge rather quickly—a trait that has both positive and negative implications. Namely, for the environments where the results are satisfactory (i.e., comparable to the state-of-the-art) this implies a reduction in computational costs. However, for the tasks where GP has not matched RL, it indicates that evolution has stagnated in a hard-to-escape local optimum, which clearly hinders the overall performance.

Reasoning further in this direction, we examine some videos of the behaviors of the graph policies and study the reward model of hopper, walker2d, and ant, to investigate the reasons underlying premature convergence. From the visual analysis, we see that the graph policies do make the agents move, yet slowly and very cautiously. This is coherent with the reward model of the environment, where (1) agents are rewarded for maintaining their stability, and (2) the episode

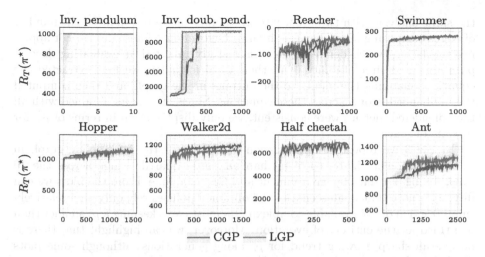

Fig. 2. Progression of the cumulative reward $R_T(\pi^\star)$ collected by the best policy in the population $\pi^\star$ at each generation in an episode of duration $T = 1000$ time steps; median and inter-quartile range across 10 independent runs.

is terminated early if an agent falls (thus preventing the collection of any additional rewards). We speculate that such a model creates a large local optimum for graph GP, where innovative policies which would require some fine tuning to become stable get swept away by evolutionary pressure.

Along this hypothesis, we experimented disabling the early termination of the episode upon the agent fall. However, the results did not show notable improvements, hinting that future work is needed to examine the phenomenon more carefully. Yet, interestingly, we found some physically unrealistic policies for the hopper, which would collect a reward as high as 30 000 by making the agent fly (video[1]). Clearly, these policies are exploiting some instability within the Brax simulator [12], though being white-boxes they could be thoroughly studied to addressing the discovered shortcoming.

5.2 Interpretability of Resulting Policies

To evaluate the interpretability of a graph policy π, we consider $\rho_c(\pi)$, which we define as the fraction of complexity employed w.r.t. the available one. In practical terms, for CGP, ρ_c is the fraction between the amount of nodes in the policy graph and the maximum possible nodes (50 in our case), whereas for LGP, ρ_c derives from the amount of program lines that contribute to the output computation divided by the total lines of the program (15 in our case).

Clearly, this is a *proxy* for interpretability, as (1) there is no clear shared definition of interpretability, and (2) interpretability can be a highly subjective notion [46]. Therefore, to ensure the general validity of our results, we repeated

---

[1] https://giorgia-nadizar.github.io/cgpax//hopper_cgp_flying.

the same analysis with two other interpretability measures: namely, the number of edges composing the policy graph, and the formula ϕ found in [45]. For applying the latter, which derives from a user study targeted at evaluating decomposability and simulatability of mathematical formulae, we first extracted the equations mapping the inputs to the outputs in the graph, and then computed the median value of ϕ across these equations. Since the results obtained with all the considered metrics were consistent, we only display them in terms of ρ_c, for reasons of space.

In Fig. 3 we report the progression of $\rho_c(\pi^\star)$ for the best policy graph in the population $\pi^\star$ in terms of median and inter-quartile range across the 10 runs, diving our data by environment and GP technique. From the plots we can distinctly note that in all scenarios the obtained policies rely only on a relatively small fraction of the available complexity, in most cases keeping a ρ_c smaller than one third for the entirety of evolution. Moreover, we can highlight that there is no overall sharp growing trend for ρ_c along generations: although some plots display a slight upward tendency, most of them stay approximately constant or even decrease with generations. This is particularly interesting to observe in comparison with Fig. 2. In fact, we discover that a sharp increase in fitness does not imply the growth of the policies, i.e., performance does not come at the cost of interpretability.

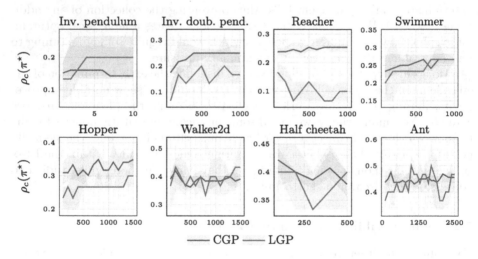

Fig. 3. Fraction of complexity employed $\rho_c(\pi^\star)$ for the best performing policy graph in the population $\pi^\star$; median and inter-quartile range across 10 independent runs.

Notably, these results naturally emerge from the representations of CGP and LGP [11,32], without any form of explicit interpretability promotion, which leads us to confirm the natural bias of these graph GP techniques towards small and interpretable graphs.

For completeness, we also repeated the optimization in a bi-objective setting—aiming at both effectiveness and interpretability—as in [44], employing

NSGA-II [8] in place of the GA described in Sect. 3.1. As before, we performed 10 runs for each scenario, although we only considered two environments: inverted double pendulum and swimmer.

We present such results in Fig. 4, where we plot the cumulative reward $R_T(\pi^\star)$ vs. the fraction of complexity $\rho_c(\pi^\star)$ for all the policies $\pi^\star$ in each of the final Pareto fronts. For context and reference, we also include the policies obtained with the standard GA. From this figure we notice that NSGA-II is able to discover simpler policies than the standard GA, though it oftentimes fails to find slightly larger but more effective ones, due to its bias towards the satisfaction of its easiest objective [29]. Thus, we can conclude that for these GP techniques, where interpretability tends to naturally emerge, it is advisable to maintain a single objective in the search.

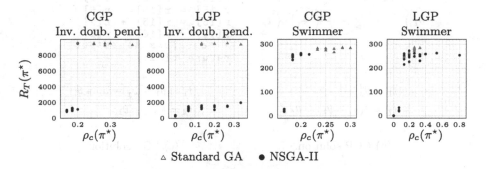

Fig. 4. Cumulative reward $R_T(\pi^\star)$ vs. fraction of complexity $\rho_c(\pi^\star)$ for each policy $\pi^\star$ in each pareto front at the end of evolution with NSGA-II. Policies found with the standard GA added for reference.

To conclude our study on the interpretability of the obtained policies, we analyze two policies for the swimmer environment. We report a schematic representation of the environment [5], together with the two policies and their conversion in equation form, in Fig. 5, and make the corresponding videos available[2].

By looking at both Figs. 5b and 5c, we can immediately perceive their simplicity and their compliance to the transparency objective in interpretability. Although in a dynamical system it is hard to achieve full understanding without considering all the equations governing it, as, e.g., those from the simulator, we can still gain some intuitions from the reported formulae. In particular, we can notice a dependency of each torque control value from the state of the opposite side of the agent, either in terms of angles or in terms of angular or linear velocities. This trait, which could not be easily detected in a black-box policy, is of paramount importance for achieving coordination of movement, and is likely a key factor for accomplishing outstanding results at the locomotion task.

---

[2] https://giorgia-nadizar.github.io/cgpax/swimmer_cgp
   https://giorgia-nadizar.github.io/cgpax/swimmer_lgp.

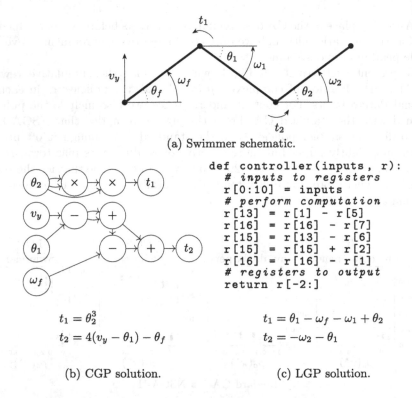

(a) Swimmer schematic.

(b) CGP solution.

(c) LGP solution.

```
def controller(inputs, r):
    # inputs to registers
    r[0:10] = inputs
    # perform computation
    r[13] = r[1] - r[5]
    r[16] = r[16] - r[7]
    r[15] = r[13] - r[6]
    r[15] = r[15] + r[2]
    r[16] = r[16] - r[1]
    # registers to output
    return r[-2:]
```

$$t_1 = \theta_2^3$$
$$t_2 = 4(v_y - \theta_1) - \theta_f$$

$$t_1 = \theta_1 - \omega_f - \omega_1 + \theta_2$$
$$t_2 = -\omega_2 - \theta_1$$

Fig. 5. Schematic representation (5a) of the swimmer environment with the observed variables $\theta_f, \theta_1, \theta_2, \omega_f, \omega_1, \omega_2, v_y$ and the control variables t_1, t_2, and the policies found with CGP (5b) and LGP (5c), shown in their original form and as formulae.

6   Concluding Remarks

In this work, we leveraged two graph-based Genetic Programming (GP) techniques, Cartesian Genetic Programming (CGP) and Linear Genetic Programming (LGP), to address several continuous control problems, the goal being that of obtaining effective yet interpretable control policies. Experimenting on eight Mujoco environments, we found graph policies that not only did perform at state-of-the-art level on a subset of tasks, but were also simple enough to be observed and directly understood. In fact, we confirmed the natural trend of both CGP and LGP to yield simple graphs, even in the absence of an explicit incentive, and found single objective optimization to be more successful in finding a good trade-off between performance and simplicity, in comparison to bi-objective optimization.

In the future, we plan to tackle some of the observed shortcomings of graph-based GP, as, e.g., premature convergence. To this end, we intend to experiment with quality-diversity optimization, novelty search, or diversity preserving mechanisms, to prevent stagnation and enable the discovery of more effective policies.

Acknowledgements. The paper is based upon work from a scholarship supported by SPECIES (http://species-society.org), the Society for the Promotion of Evolutionary Computation in Europe and its Surroundings. This study was carried out within the PNRR research activities of the consortium iNEST (Interconnected North-Est Innovation Ecosystem) funded by the European Union Next-GenerationEU (Piano Nazionale di Ripresa e Resilienza (PNRR) - Missione 4 Componente 2, Investimento 1.5 - D.D. 1058 23/06/2022, ECS_00000043).

References

1. Adadi, A., Berrada, M.: Peeking inside the black-box: a survey on explainable artificial intelligence (XAI). IEEE Access **6**, 52138–52160 (2018)
2. Amaral, R., Ianta, A., Bayer, C., Smith, R.J., Heywood, M.I.: Benchmarking genetic programming in a multi-action reinforcement learning locomotion task. In: Proceedings of the Genetic and Evolutionary Computation Conference Companion, pp. 522–525 (2022)
3. Bradbury, J., et al.: Jax: composable transformations of python+ numpy programs (2018)
4. Brameier, M., Banzhaf, W., Banzhaf, W.: Linear Genetic Programming, vol. 1. Springer, New York (2007). https://doi.org/10.1007/978-0-387-31030-5
5. Coulom, R.: Reinforcement learning using neural networks, with applications to motor control. Ph.D. thesis, Institut National Polytechnique de Grenoble-INPG (2002)
6. Custode, L.L., Iacca, G.: Evolutionary learning of interpretable decision trees. arXiv preprint arXiv:2012.07723 (2020)
7. Custode, L.L., Iacca, G.: Interpretable pipelines with evolutionary optimized modules for reinforcement learning tasks with visual inputs. In: Proceedings of the Genetic and Evolutionary Computation Conference Companion, pp. 224–227 (2022)
8. Deb, K., Pratap, A., Agarwal, S., Meyarivan, T.: A fast and elitist multiobjective genetic algorithm: Nsga-ii. IEEE Trans. Evol. Comput. **6**(2), 182–197 (2002)
9. Ferigo, A., Custode, L.L., Iacca, G.: Quality diversity evolutionary learning of decision trees. arXiv preprint arXiv:2208.12758 (2022)
10. Ferigo, A., Custode, L.L., Iacca, G.: Quality-diversity optimization of decision trees for interpretable reinforcement learning. Neural Comput. Appl. 1–12 (2023)
11. Françoso Dal Piccol Sotto, L., Kaufmann, P., Atkinson, T., Kalkreuth, R., Porto Basgalupp, M.: Graph representations in genetic programming. Genet. Program. Evolvable Mach. **22**(4), 607–636 (2021)
12. Freeman, C.D., Frey, E., Raichuk, A., Girgin, S., Mordatch, I., Bachem, O.: Brax-a differentiable physics engine for large scale rigid body simulation. arXiv preprint arXiv:2106.13281 (2021)
13. Glanois, C., Weng, P., Zimmer, M., Li, D., Yang, T., Hao, J., Liu, W.: A survey on interpretable reinforcement learning. arXiv preprint arXiv:2112.13112 (2021)
14. Glass, A., McGuinness, D.L., Wolverton, M.: Toward establishing trust in adaptive agents. In: Proceedings of the 13th International Conference on Intelligent User Interfaces, pp. 227–236 (2008)
15. Guidotti, R., Monreale, A., Ruggieri, S., Turini, F., Giannotti, F., Pedreschi, D.: A survey of methods for explaining black box models. ACM Comput. Surv. (CSUR) **51**(5), 1–42 (2018)

16. Haarnoja, T., Zhou, A., Abbeel, P., Levine, S.: Soft actor-critic: off-policy maximum entropy deep reinforcement learning with a stochastic actor. In: International Conference on Machine Learning, pp. 1861–1870, PMLR (2018)

17. Hein, D., Udluft, S., Runkler, T.A.: Interpretable policies for reinforcement learning by genetic programming. Eng. Appl. Artif. Intell. **76**, 158–169 (2018)

18. Henderson, P., Islam, R., Bachman, P., Pineau, J., Precup, D., Meger, D.: Deep reinforcement learning that matters. In: Proceedings of the AAAI Conference on Artificial Intelligence, vol. 32 (2018)

19. Kantschik, W., Banzhaf, W.: Linear-graph GP - a new GP structure. In: Foster, J.A., Lutton, E., Miller, J., Ryan, C., Tettamanzi, A. (eds.) EuroGP 2002. LNCS, vol. 2278, pp. 83–92. Springer, Heidelberg (2002). https://doi.org/10.1007/3-540-45984-7_8

20. Kaufmann, E., Bauersfeld, L., Loquercio, A., Müller, M., Koltun, V., Scaramuzza, D.: Champion-level drone racing using deep reinforcement learning. Nature **620**(7976), 982–987 (2023)

21. Kelly, S., Heywood, M.I.: Emergent tangled graph representations for Atari game playing agents. In: McDermott, J., Castelli, M., Sekanina, L., Haasdijk, E., García-Sánchez, P. (eds.) EuroGP 2017. LNCS, vol. 10196, pp. 64–79. Springer, Cham (2017). https://doi.org/10.1007/978-3-319-55696-3_5

22. Kelly, S., Heywood, M.I.: Multi-task learning in atari video games with emergent tangled program graphs. In: Proceedings of the Genetic and Evolutionary Computation Conference, pp. 195–202 (2017)

23. Kelly, S., et al.: Discovering adaptable symbolic algorithms from scratch. arXiv preprint arXiv:2307.16890 (2023)

24. Kelly, S., Voegerl, T., Banzhaf, W., Gondro, C.: Evolving hierarchical memory-prediction machines in multi-task reinforcement learning. Genet. Program Evolvable Mach. **22**, 573–605 (2021)

25. Koza, J.R.: Genetic programming as a means for programming computers by natural selection. Stat. Comput. **4**(2), 87–112 (1994)

26. Koza, J.R., Rice, J.P.: Automatic programming of robots using genetic programming. In: AAAI, vol. 92, pp. 194–207 (1992)

27. Landajuela, M., et al.: Discovering symbolic policies with deep reinforcement learning. In: International Conference on Machine Learning, pp. 5979–5989, PMLR (2021)

28. Lipton, Z.C.: The mythos of model interpretability: in machine learning, the concept of interpretability is both important and slippery. Queue **16**(3), 31–57 (2018)

29. Liu, D., Virgolin, M., Alderliesten, T., Bosman, P.A.: Evolvability degeneration in multi-objective genetic programming for symbolic regression. In: Proceedings of the Genetic and Evolutionary Computation Conference, pp. 973–981 (2022)

30. Machado, M.C., Bellemare, M.G., Talvitie, E., Veness, J., Hausknecht, M., Bowling, M.: Revisiting the arcade learning environment: evaluation protocols and open problems for general agents. J. Artif. Intell. Res. **61**, 523–562 (2018)

31. Medvet, E., Nadizar, G.: GP for continuous control: teacher or learner? The case of simulated modular soft robots. In: Winkler, S., Trujillo, L., Ofria, C., Hu, T. (eds.) Genetic Programming Theory and Practice XX. Genetic and Evolutionary Computation, Springer, Singapore (2023). https://doi.org/10.1007/978-981-99-8413-8_11

32. Miller, J.F.: Cartesian genetic programming: its status and future. Genet. Program Evolvable Mach. **21**, 129–168 (2020)

33. Miller, J.F., Thomson, P.: Cartesian genetic programming. In: Poli, R., Banzhaf, W., Langdon, W.B., Miller, J., Nordin, P., Fogarty, T.C. (eds.) EuroGP 2000.

LNCS, vol. 1802, pp. 121–132. Springer, Heidelberg (2000). https://doi.org/10. 1007/978-3-540-46239-2_9

34. Nadizar, G., Rovito, L., De Lorenzo, A., Medvet, E., Virgolin, M.: An analysis of the ingredients for learning interpretable symbolic regression models with human-in-the-loop and genetic programming. ACM Tran. Evol. Learn. (2024)
35. Puiutta, E., Veith, E.M.S.P.: Explainable reinforcement learning: a survey. In: Holzinger, A., Kieseberg, P., Tjoa, A.M., Weippl, E. (eds.) CD-MAKE 2020. LNCS, vol. 12279, pp. 77–95. Springer, Cham (2020). https://doi.org/10.1007/978-3-030-57321-8_5
36. Puterman, M.L.: Markov Decision Processes: Discrete Stochastic Dynamic Programming. Wiley, Hoboken (2014)
37. Rudin, C.: Stop explaining black box machine learning models for high stakes decisions and use interpretable models instead. Nat. Mach. Intell. **1**(5), 206–215 (2019)
38. Salvato, E., Fenu, G., Medvet, E., Pellegrino, F.A.: Crossing the reality gap: a survey on sim-to-real transferability of robot controllers in reinforcement learning. IEEE Access **9**, 153171–153187 (2021)
39. Schulman, J., Wolski, F., Dhariwal, P., Radford, A., Klimov, O.: Proximal policy optimization algorithms. arXiv preprint arXiv:1707.06347 (2017)
40. Sigaud, O., Stulp, F.: Policy search in continuous action domains: an overview. Neural Netw. **113**, 28–40 (2019)
41. Silver, D., et al.: Mastering the game of go with deep neural networks and tree search. Nature **529**(7587), 484–489 (2016)
42. Todorov, E., Erez, T., Tassa, Y.: Mujoco: a physics engine for model-based control. In: 2012 IEEE/RSJ International Conference on Intelligent Robots and Systems, pp. 5026–5033. IEEE (2012)
43. Verma, A., Murali, V., Singh, R., Kohli, P., Chaudhuri, S.: Programmatically interpretable reinforcement learning. In: International Conference on Machine Learning, pp. 5045–5054. PMLR (2018)
44. Videau, M., Leite, A., Teytaud, O., Schoenauer, M.: Multi-objective genetic programming for explainable reinforcement learning. In: Medvet, E., Pappa, G., Xue, B. (eds.) EuroGP 2022. LNCS, vol. 13223, pp. 278–293. Springer, Cham (2022). https://doi.org/10.1007/978-3-031-02056-8_18
45. Virgolin, M., De Lorenzo, A., Medvet, E., Randone, F.: Learning a formula of interpretability to learn interpretable formulas. In: Bäck, T., et al. (eds.) PPSN 2020. LNCS, vol. 12270, pp. 79–93. Springer, Cham (2020). https://doi.org/10. 1007/978-3-030-58115-2_6
46. Virgolin, M., De Lorenzo, A., Randone, F., Medvet, E., Wahde, M.: Model learning with personalized interpretability estimation (ml-pie). In: Proceedings of the Genetic and Evolutionary Computation Conference Companion, pp. 1355–1364 (2021)
47. Wells, L., Bednarz, T.: Explainable AI and reinforcement learning-a systematic review of current approaches and trends. Front. Artif. Intell. **4**, 550030 (2021)
48. Wilson, D.G., Cussat-Blanc, S., Luga, H., Miller, J.F.: Evolving simple programs for playing Atari games. In: Proceedings of the Genetic and Evolutionary Computation Conference, pp. 229–236 (2018)
49. Wilson, D.G., Miller, J.F., Cussat-Blanc, S., Luga, H.: Positional cartesian genetic programming. arXiv preprint arXiv:1810.04119 (2018)
50. Zhou, R., Hu, T.: Evolutionary approaches to explainable machine learning. arXiv preprint arXiv:2306.14786 (2023)

DALex: Lexicase-Like Selection
via Diverse Aggregation

Andrew Ni[1]([✉])(iD), Li Ding[2](iD), and Lee Spector[1,2](iD)

[1] Amherst College, Amherst, MA 01002, USA
{ani24,lspector}@amherst.edu
[2] University of Massachusetts Amherst, Amherst, MA 01003, USA
liding@umass.edu

Abstract. Lexicase selection has been shown to provide advantages over other selection algorithms in several areas of evolutionary computation and machine learning. In its standard form, lexicase selection filters a population or other collection based on randomly ordered training cases that are considered one at a time. This iterated filtering process can be time-consuming, particularly in settings with large numbers of training cases, including many symbolic regression and deep learning applications. In this paper, we propose a new method that is nearly equivalent to lexicase selection in terms of the individuals that it selects, but which does so in significantly less time. The new method, called DALex (for Diversely Aggregated Lexicase selection), selects the best individual with respect to a randomly weighted sum of training case errors. This allows us to formulate the core computation required for selection as matrix multiplication instead of recursive loops of comparisons, which in turn allows us to take advantage of optimized and parallel algorithms designed for matrix multiplication for speedup. Furthermore, we show that we can interpolate between the behavior of lexicase selection and its "relaxed" variants, such as epsilon and batch lexicase selection, by adjusting a single hyperparameter, named "particularity pressure," which represents the importance granted to each individual training case. Results on program synthesis, deep learning, symbolic regression, and learning classifier systems demonstrate that DALex achieves significant speedups over lexicase selection and its relaxed variants while maintaining almost identical problem-solving performance. Under a fixed computational budget, these savings free up resources that can be directed towards increasing population size or the number of generations, enabling the potential for solving more difficult problems.

Keywords: Lexicase Selection · Learning Classifier Systems · Genetic Programming · Symbolic Regression · Deep Learning

Supported by Amherst College and members of the PUSH lab.

1 Introduction

Genetic algorithms [26] is a subfield of evolutionary computation that uses concepts from biological evolution—random variation and fitness-based survival—to solve a wide array of difficult problems. Genetic programming (GP) is a subfield of genetic algorithms that evolves programs or functions that take some inputs and produce some outputs. Common subfields of GP include software synthesis, in which a population of programs evolves to satisfy user-defined training cases, and symbolic regression (SR), in which a population of mathematical expressions evolves to fit a regression dataset. Central to these genetic algorithms is the concept of parent selection, in which members of the current population with desirable characteristics are chosen as the starting point from which to create the next generation of individuals. Lexicase selection [24] and epsilon lexicase selection [32] are state-of-the-art selection algorithms developed for the discrete-error domain and the continuous-error domain, respectively. The key idea behind lexicase selection is that it can be helpful to disaggregate the fitness function, selecting parents by considering training cases one at a time in a random order, all the while filtering out individuals which are not elite relative to the other remaining individuals on the current training case. This selection method has been shown to maintain beneficial diversity in evolved populations [17, 36], especially in terms of "specialists": individuals which may have high total error but have very low errors on a subset of the training cases [19, 20]. However, due to its iterative nature, lexicase selection can often take a long time, and there is not an obvious way to take advantage of parallelism or single-instruction-multiple-data (SIMD) architectures to speed up its runtime.

In this work, we reexamine the lexicase intuition behind disaggregating the fitness function and we develop an efficient selection method based on randomly aggregating the fitness function at each individual selection event. Specifically, we quantify the idea that training cases occurring earlier in a lexicase ordering exert greater selection pressure by taking a randomly weighted average of each individual's error vector. We formulate this selection mechanism as a matrix multiplication, which allows us to take advantage of modern advancements in matrix multiplication algorithms and SIMD architectures to achieve significantly faster runtime compared to lexicase selection.

This paper is organized as follows: In Sect. 2, we give a brief overview of lexicase selection and its variants. In Sect. 3, we describe the Diversely Aggregated Lexicase Selection (DALex) algorithm and give a brief theoretical treatment of its properties. In Sect. 4, we describe our experiments and results in three popular domains in which lexicase selection has seen demonstrated success. Empirical results show that DALex replicates the problem-solving success of lexicase selection and its most successful variants while offering significantly reduced runtime.

2 Background and Related Work

Lexicase selection is a parent selection method that assesses individuals based on each training case in turn, instead of constructing a single scalar fitness value for each individual [24]. During each individual selection event, the training cases are randomly shuffled. For each training case in the order determined by the random shuffle, candidate individuals in the population that are not "elite" with respect to that training case, i.e., have an error greater than the minimum error on that training case among the remaining candidates, are filtered out. If at any point only a single individual remains, then that individual is selected. If all of the training cases are exhausted, then a random individual is selected from those individuals still remaining. In other words, lexicase selection chooses an individual based on the first training case, using the rest of the cases in order to break ties. This mechanism has been shown to better maintain population diversity than methods based on aggregated fitness measures [17], which has been shown to improve problem-solving performance [15]. Lexicase selection and its variants have been successfully applied in many diverse problem domains beyond software synthesis, such as learning classifier systems [1], symbolic regression [32], SAT solvers [35], deep learning [9], and evolutionary robotics [44].

In continuous-valued domains like SR, it is often unlikely for more than one individual to share the minimum error on a training case. Therefore, the selection pressure exerted by each training case will be too large, and lexicase selection will not proceed beyond one training case. To combat this, the epsilon-lexicase selection method was developed as a relaxation of lexicase selection [32]. Instead of requiring individuals to have the minimum error out of the current candidates on a training case, epsilon-lexicase selection filters out individuals with errors exceeding an epsilon threshold above the minimum error. This epsilon threshold is adaptively determined by the population dynamics, specifically as the median of the absolute deviations from the median error on each training case.

Batch-lexicase selection [1, 34, 40] is another lexicase relaxation with improved performance on noisy datasets. Instead of considering each training case in turn, batch-lexicase selection considers groups of training cases, filtering out individuals whose average accuracy on that group is lower than some threshold. Batch-lexicase selection has been successfully applied to the field of Learning Classifier Systems (LCS) in Aenugu et al. [1], showing improved performance on noisy datasets compared to lexicase selection.

There have been many attempts to speed up lexicase selection. Dolson et al. [10] showed that the exact calculation of lexicase selection probabilities is NP-hard. However, Ding et al. [8] are able to calculate an approximate probability distribution from which to sample individuals. They show that their selection method achieves significantly faster runtime while achieving similar selection frequencies compared to lexicase selection. Ding et al. [7] also propose to use a weighted shuffle of training cases so that more difficult cases are considered first. They show that this technique can significantly reduce the number of training cases considered while suffering minimal degradation of problem-solving power. Batch tournament selection (BTS) [34] orders the training cases by difficulty, then groups them into batches. BTS then performs tournament selection once per batch of cases, using the average error on that batch as the fitness function.

They show that BTS achieves similar problem-solving performance to epsilon lexicase selection on SR datasets with significantly faster runtime.

Furthermore, many selection methods based on aggregated fitness measures have been proposed for multiobjective genetic algorithms. Fitness sharing is a selection method that attempts to maintain diversity by penalizing individuals for being too close to other individuals in the population with respect to a user-defined distance metric [6]. On the other hand, implicit fitness sharing (IFS) does not require a distance metric, and instead scales the reward for each training case by the rest of the population's performance on that training case before computing the aggregated fitness function [33]. Historically assessed hardness (HAH) is another form of fitness sharing in which errors on each training case are scaled by the population's historical success rate on that problem [28]. In addition to using a distance metric, NSGA-II sorts individuals into pareto fronts and conducts tournament selection using nondomination rank and local crowding factor as fitness values [5].

This work lies at the intersection of the lexicase variants mentioned above but takes a different direction compared to the speedup methods developed so far. For each individual selection event, we sample a random weighting of training cases. Then, for each individual, we compute a weighted average of the training case errors and select the individual with the lowest error with respect to the weighted average. In contrast to IFS or HAH, we use weights that are randomly chosen, span many more orders of magnitude, and are resampled at each selection event instead of at each generation. In contrast to NSGA-II, this method only selects non-dominated individuals because all of the training case weights are positive after the softmax operation. However, like epsilon and batch lexicase selection, and in contrast to lexicase selection, this method may select individuals outside of the "Pareto boundary" as defined in La Cava et al. [30]. We formulate this selection method as a vectorized matrix multiplication, selecting all individuals in a single, batched selection event. This allows us to take advantage of modern SIMD architecture and algorithmic advances in matrix multiplication to achieve significantly faster runtime compared to lexicase selection.

3 Diversely Aggregated Lexicase Selection

3.1 Description

Diversely Aggregated Lexicase Selection (DALex) operates by selecting the individual with the lowest average error with respect to a randomly sampled set of weights. It is parameterized by the shape and scale of the distribution from which these weights are sampled. In most of our experiments, we fix the shape to that of the normal distribution and vary the scale by changing the standard deviation of the distribution. For an ablation experiment exploring the effect of different distributions on the performance of DALex, see the appendix[1] We start with a population of n individuals, where each individual is evaluated on m training

---

[1] https://arxiv.org/abs/2401.12424.

cases. We sample an importance score for each training case from the distribution $\mathcal{N}(0, \text{std})$ where the standard deviation std is a tunable hyperparameter which we call the *particularity pressure*. Generally speaking, a training case's importance score quantifies how many times more important it is compared to the rest of the training cases. To obtain the training case weights, we softmax the importance scores. This allows us to turn differences between the values of importance scores into differences between the magnitudes of the training case weights. For each individual, we compute its average weighted error as the dot product of the training case weight vector and the individual's error vector. Finally, we select the individual with the lowest average weighted error.

In practice, we conduct all individual selection events in one batched selection event, combining multiple weight vectors into a weight matrix and using matrix multiplication in place of the dot product. We report selection runtimes in terms of this batched selection event, i.e. the time taken to go from a population of n individuals to the indices of the n selected parents for the next generation. We also perform an initial "pre-selection" step that differs slightly from other lexicase selection implementations due to the batched nature of our method: we group individuals into equivalence classes based on their error vectors, select equivalence classes using DALex, and then choose a random individual from each selected class [16].

For a population of n individuals evaluated on m training cases, there will be $k \leq n$ distinct equivalence classes, so the algorithm multiplies a $k \times m$ error matrix with a transposed $n \times m$ weight matrix, giving an asymptotic runtime of $O(m^2 n)$, which is the same as that of lexicase selection. However, due to advances in the theory of matrix multiplication, the runtime of DALex can be reduced to $O(n^{\omega(\log_n m)})$ using methods such as the Coppersmith-Winograd algorithm [4].

3.2 Intuition

For a better understanding of DALex, we provide two intuitions linking it to lexicase selection.

First, we argue that DALex prompts us to reconsider a central tenet of lexicase selection. The success of lexicase selection is commonly attributed to its disaggregation of the fitness function. As the reasoning goes, aggregated selection methods like tournament or fitness-proportional selection are unable to maintain effective problem solving diversity the same way lexicase selection does. To rationalize our results on DALex, we instead propose the slightly different view that the discrepancy in problem-solving performance is due to the single aggregation event preceding these selection methods, which causes a loss of information. Because lexicase selection utilizes the entire high-dimensional error vector instead of simply the average or sum, it is able to more accurately identify promising individuals. Since DALex uses many diverse aggregation events, it is also able to fully utilize the information contained in the individuals' error vectors, and is therefore able to replicate the success of lexicase selection. Even though each individual selection event operates on a single-dimensional total error, diversity arises from different random weights promoting different individuals in different selection events.

Second, we draw a correspondence between the random orderings in lexicase selection and the random weightings in DALex. Given a random ordering of training cases in lexicase selection, the cases occurring earlier in the ordering exert more control over which individual is chosen at that selection event, i.e. have higher selection pressure. The first training case is paramount, and only if multiple individuals are equally good at the first case do we consider the second case, and so on. In DALex, the importance scores or training case weights assign explicit values to this notion of selection pressure. Given training case weights of [1.0, 0.01, 0.1] for example, the first case would be ten times as important as the third case, which in turn is ten times as important as the second case. In the limit of infinite particularity pressure, the differences in importance scores tend towards infinity, so the difference in magnitudes of training case weights tends to infinity, and we recover lexicase selection. As the number of training cases increases or the range of magnitudes spanned by the errors on each case increases, the particularity pressure needed to replicate lexicase selection increases. Empirically, we are able to replicate lexicase selection using modest particularity pressures such as 20.

This interpretation also explains why we call the standard deviation hyperparameter the particularity pressure. When importance weights are sampled with a high standard deviation, the most important training case will be much more important than the second most, and so on, just like in lexicase selection. As the standard deviation decreases, the importance of a single training case decreases. In other words, higher particularity pressures correspond to increasingly lexicase-like selection dynamics [42]. We find that we can achieve similar successes to relaxed forms of lexicase selection such as batch-lexicase or epsilon-lexicase selection simply by choosing an appropriate particularity pressure. In this sense, DALex unifies the diverse selection methods lexicase, epsilon-lexicase, and batch-lexicase selection into a single selection method. It is important to note that DALex's only theoretical guarantee is that of asymptotically lexicase-like selection with infinite particularity pressure. As decreasing the particularity pressure is a different way of relaxation compared to those proposed for epsilon-lexicase and batch-lexicase selection, we do not expect to be able to replicate the selection dynamics of these lexicase variants to an arbitrary degree of accuracy. However, empirical results show that varying the particularity pressure gives enough flexibility to replicate the successes of these relaxed lexicase variants on the problem domains for which they were designed.

3.3   Modifications

For domains amenable to a relaxed lexicase variant such as epsilon-lexicase or batch-lexicase selection, we first standardize the population errors on each training case to have zero mean and unit variance. While not necessary, we find that this normalization step helps DALex perform well across many problems with a single hyperparameter setting.

Of the problem domains we study, Learning Classifier Systems (LCS) [46] has a significantly different structure and therefore requires further modification. Since an LCS individual represents a rule that matches a subset of the training cases, each individual will only have errors defined on a subset of the training cases. For these problems, we let the individual have error 0 on cases for which it is undefined. We also use the individual's support vector, which has a 1 for each training case on which the individual is defined, and a 0 otherwise. To compute the individual's average error, we take the dot product of the error vector with the weight vector and then normalize by dividing this value by the dot product of the support vector with the weight vector.

Algorithm 1: Diversely Aggregated Lexicase Selection, batched selection event

---

Data:

- $E = \{e_{i,j}\}$ the error value of individual i on training case j, or 0 if individual i is not defined on training case j
- $N = \{s_{i,j}\}$ the support of individual i on training case j, which equals 1 if individual i is defined on case j, otherwise 0
- n the number of selection events, m the number of training cases
- `particularity_pressure`, the standard deviation of the sampled weights
- `Relaxed`, whether to simulate a relaxed version of lexicase selection.

Result:

- `idx`, the n indices of the selected individuals to be the parents of the next generation.

if Relaxed **then**

$\quad \Big|\quad E \leftarrow \frac{E - \texttt{mean}(E,\texttt{axis}=0)}{\texttt{std}(E,\texttt{axis}=0)}$ The standardized error matrix

end

$I \leftarrow \mathcal{N}(0, \texttt{particularity\_pressure})$ The importance scores, an i.i.d gaussian matrix of size $[n, m]$

$W \leftarrow \texttt{softmax}(I, \texttt{axis} = 1)$ The training case weights

$F \leftarrow \frac{EW^T}{SW^T}$ The normalized, weighted, average error for each individual

$\texttt{idx} \leftarrow \texttt{argmin}(F, \texttt{axis} = 0)$ lowest error individuals w.r.t the weights W

return idx

In problems where individuals are defined on all training cases, the support is 1 everywhere, so the normalization constant SW^T is simply the sum of the training case weights, which is 1 due to the softmax operation. Under this condition, our algorithm degenerates to the simpler case considered above. The complete algorithm pseudocode, with the augmentations described above to accommodate for LCS and SR problems, is described in Algorithm 1. In our pseudocode, we define arithmetic operations as the vectorized element-wise operations used in libraries like `numpy`, and batch together the n importance score vectors for the n individual selection events into an $n \times m$ matrix.

4   Experiments and Results

4.1   CBGP

We first compare DALex to lexicase selection on program synthesis problems, the class of problem for which lexicase selection was first developed. While we use a particularity pressure of 20 for these experiments to illustrate the robustness of DALex, we recommend setting the particularity pressure as large as possible within the range of floating point precision, or around 200, for problems suitable for lexicase selection. As DALex can exactly simulate lexicase selection in the limit of infinite particularity pressure, we also assess the ability of DALex to replicate the lexicase selection probability distribution with a moderate particularity pressure. To situate our results with respect to other lexicase approximations we also compare to plexicase selection[2] [8]. For plexicase selection, we use the hyperparameter setting $\alpha = 1$. We use the Code-Building Genetic Programming (CBGP) system developed by Helmuth et al. [37] and test on a suite of problems from the PSB1 Benchmark [21] on which lexicase selection has shown good performance. For an overview on CBGP, see the appendix. We follow the default settings in CBGP, evolving 1000 individuals for 300 generations and using UMAD [18] with a rate of 0.09 as the sole variation operator. We present results in both the full data and downsampled paradigms. In the full data paradigm, all individuals are evaluated on all training cases. In the downsampled paradigm, a subset of the training cases is sampled at each generation and used to evaluate individuals in that generation. Downsampling has been shown to benefit the performance of Genetic Programming as we can increase the number of generations and/or individuals in the population for a given total computational budget [22,23,25]. Additional downsampling methods utilizing information from the population to choose the downsampled cases have been proposed [2], but we only study the randomly downsampled paradigm. In our experiments, we use a downsampling rate of 25%. The specific problems we study are `Compare String Lengths` (CSL), `Median`, `Number IO`, `Replace Space with Newline` (RSWN), `Smallest`, `Vector Average`, and `Negative to Zero` (NTZ).

Following the recommendation of the PSB1 benchmark, we report problem-solving performance as number of successful runs out of 100. We define a successful run as one which produces an individual with zero error on both the training cases and the unseen testing cases.[3]

Table 1 compares the success rates of DALex, plexicase, and lexicase selection on six benchmark problems under the full and downsampled paradigms. No results were significantly different between the selection methods, and DALex achieves very similar success rates to lexicase selection. For our ablation experiments on the distribution of the sampled importance scores, see the appendix. Additionally, we find that DALex runs faster than lexicase selection but slower than plexicase selection. Detailed results can be found in the appendix.

---

[2] Fixed a bug in the downsampling implementation in the released version of [8].

[3] Due to specific quirks of the code-building system, it is very difficult for CBGP to generalize successfully on `Compare String Lengths`.

Table 1. Success rates of GP runs on six benchmark problems in the full data and downsampled paradigms. The success rates across the board are very similar, and no results were significantly different between the selection methods. We perform chi-squared tests following [8] and underline results that were significantly worse than lexicase selection with ($p < 0.05$). No results were significantly better than lexicase selection.

Problem	Success Rate			Downsampled Success Rate		
	Lexicase	*DALex*	*Plexicase* ($\alpha = 1$)	*Lexicase*	*DALex*	*Plexicase* ($\alpha = 1$)
CSL[a]	0	0	0	0	0	0
Median	91	91	<u>83</u>	100	100	60
Number IO	99	99	100	100	100	100
RSWN	12	15	6	66	68	<u>50</u>
Smallest	100	100	100	100	100	100
Vector Average	100	100	99	100	100	100
NTZ	78	83	80	99	100	100

[a] Due to specific quirks, `Compare String Lengths` is systemically difficult for CBGP to solve.

To compare the selection dynamics of the three methods, we solve the same PSB1 problems using lexicase selection, and at every generation we sample 50,000 individuals using each selection method. From this we build up the empirical probability distributions $\{p_{it}\}$ the probability of selecting individual i at generation t via lexicase selection, $\{p'_{it}\}$ the probability of selecting individual i at generation t via DALex, and $\{p''_{it}\}$ the probability of selecting individual i at generation t via plexicase selection. To quantify the differences in probability distributions between lexicase selection and its approximations, we use the Jensen-Shannon divergence metric

$$D_t = \frac{1}{2}\left[\sum_i p_{it} \log\left(\frac{2p_{it}}{p_{it} + q_{it}}\right) + \sum_i q_{it} \log\left(\frac{2q_{it}}{p_{it} + q_{it}}\right)\right]$$

Where q_{it} is p'_{it} or p''_{it} for DALex and plexicase selection, respectively. The lower the JS-divergence of the DALex/plexicase selection probability distribution from the lexicase selection probability distribution, the better the approximation to lexicase selection. Furthermore, for each run in which lexicase selection produces a generalizing individual, we track that individual's lineage to find its ancestor a_t in each generation t, of which there is only one per generation since we only use mutation operators. For each of these individuals we compute the probability ratio $r_t = \frac{q_{a_t t}}{p_{a_t t}}$ quantifying how much more likely DALex and plexicase selection are to select the successful lineage than lexicase selection. We report the average of these two metrics over all generations and all runs.

Figure 1 shows the results for these two metrics on the problems studied. In almost all problems, the probability of selecting the successful lineage is almost the same between DALex and lexicase selection. In contrast, the probability ratio of plexicase and lexicase selection often deviates much more from 1, indi-

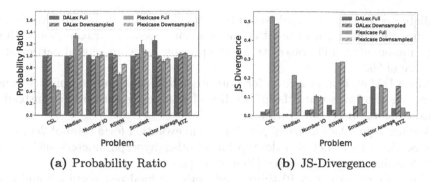

(a) Probability Ratio   (b) JS-Divergence

Fig. 1. Fidelity of the DALex and plexicase approximations to lexicase selection on CBGP problems. Error bars show the bootstrapped 95% confidence intervals. The ratio of selecting the successful lineage via DALex versus the probability using lexicase is very close to 1 in 6 out of 7 problems. The Jensen-Shannon divergence of the DALex selection probability distribution from the lexicase distribution is close to 0 in 5 out of 7 problems.

cating a less faithful approximation. Furthermore, the Jensen-Shannon divergences between the DALex and lexicase selection probability distributions are often very close to zero and much smaller than the corresponding divergences of plexicase selection from lexicase selection, indicating that DALex gives a close approximation of lexicase selection dynamics. As mentioned before, we expect increasing the value of the standard deviation to improve the fidelity of the DALex approximation, bringing the DALex JS divergences even lower and the DALex probability ratios even closer to 1.

4.2   Image Classification

To test the ability of DALex to mimic lexicase selection even with many training cases, we compare DALex against lexicase selection as the selection method in gradient lexicase selection. We use the popular VGG16 [39] and ResNet18 [14] neural networks on the CIFAR10 [29] dataset. This dataset is comprised of 32×32 bit RGB images distributed evenly across 10 classes. There are 50,000 training instances and 10,000 test instances. As shown in Ding et al. [7], lexicase selection often uses up to the entire training set to perform selection, especially towards the end of the genetic algorithm.

Gradient lexicase selection [9] is a hybrid of lexicase selection with deep learning, in which a population of deep neural networks (DNN) is evolved to fit an image classification dataset. In gradient lexicase selection, stochastic gradient descent on a subset of the training cases is used as the mutation operator. Lexicase selection proceeds by evaluating the population of DNNs on one training case at a time and keeping only the DNNs that correctly predict the training case's label. In contrast, DALex evaluates each individual on the entire training set, assigns to each training case an error value of 0 if the case was correctly predicted or 1 otherwise, and computes a weighted sum of the error values. As this domain is intended to test the limits of our lexicase approximation with its large number of training cases, we use the previously recommended particularity pressure of 200.

Since DALex is appropriate for both discrete-valued and continuous-valued errors, we also investigate the possibility of using cross-entropy loss instead of accuracy as the basis of selection.

We use a population size of $p = 4$ and run evolution for a total of $200(p + 1)$ epochs. For a more detailed description of other hyperparameters such as the SGD learning rate schedule, see Ding et al. [9].

Each algorithm is run 10 times, and both the final test accuracy and algorithm runtime are recorded. All runs were done on a single Nvidia A100 gpu. We report the mean and standard deviation of these two metrics in Table 2 the form mean ± std.

Table 2. Performance of three selection methods and two architectures on the CIFAR10 dataset. Vanilla SGD is also included as a baseline. DALex acheives similar test accuracy and lower runtime compared to lexicase selection.

Selector	VGG16		ResNet18	
	Test Accuracy	*Runtime* (h)	*Test Accuracy*	*Runtime* (h)
Vanilla SGD	92.8 ± 0.2	0.962 ± 0.003	93.8 ± 0.1	1.003 ± 0.006
Lexicase	93.7 ± 0.1	11.1 ± 0.3	95.1 ± 0.2	29 ± 1
DALex accuracy (std = 200)	93.7 ± 0.2	8.7 ± 0.1	95.3 ± 0.1	9.1 ± 0.1
DALex losses (std = 200)	93.7 ± 0.2	8.66 ± 0.02	95.2 ± 0.2	9.2 ± 0.1

We find that DALex performs just as well as lexicase selection whether selecting on cross-entropy losses or accuracy, and additionally runs much faster. This is likely due to the fact that lexicase selection evaluates individuals on each training case with a batch size of 1. The small batch size causes too much communication overhead between the cpu and gpu, resulting in a slower runtime. Furthermore, as training progresses and the models approach 100% accuracy on the training set, lexicase selection begins to exhibit its worst-case runtime. As the ResNet18 model reaches a higher training set accuracy than the VGG16 model, lexicase selection takes more than twice as long to run. With DALex, however, both models have consistent and much lower runtimes.

4.3   SRBench

To examine the ability of DALex to replicate epsilon lexicase-like behavior, we use the SRBench [31] benchmark suite for symbolic regression. Specifically, we compare DALex, plexicase, and epsilon lexicase selection as the selection method for the gplearn [45] framework and train on 20 subsampled datasets from the Penn Machine Learning Benchmark (PMLB) [38] suite with sizes ranging from 50 to 40,000 instances. As recommended by SRBench, we repeat each experiment 10 times with different random seeds, and take the median of each statistic over these 10 experiments. For these experiments, we use the pre-tuned settings of 1000 individuals per generation, 500 generations, and functions drawn from the set $\{+, -, *, /, \log, \operatorname{sqrt}, \sin, \cos\}$. We also use the default parsimony coefficient of 0.001, which adds a model size-based penalty to the error of each test case. We choose the DALex particularity pressure to be 3, as determined from a preliminary hyperparameter search.

The statistics we examine in this paper, following the SRBench defaults, are test R^2, training time, and model complexity. Unlike in program synthesis, we find that the runtime of epsilon lexicase dominates the algorithm runtime, so we expect the choice of selection method to significantly impact the overall runtime. We report the median of these statistics over all 20 problems, using bootstrapping to obtain a 95% confidence interval.

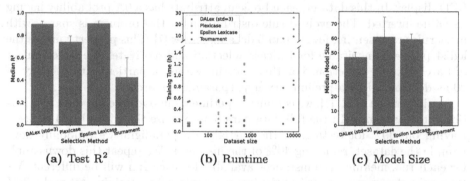

(a) Test R^2 (b) Runtime   (c) Model Size

Fig. 2. Performance of three selection methods on 20 downsampled problems from the PMLB repository. DALex has a similar test R^2 to epsilon lexicase while having a lower runtime and generating more concise models.

Figure 2 shows the results for the 20 selected problems. We find that DALex gives a very similar test R^2 statistic to epsilon lexicase selection while generating significantly more concise models and taking much less time to train. Of the problems we studied, epsilon lexicase selection has a much higher runtime on the 344_mv problem than on all other problems. We hypothesize that this is due to epsilon lexicase selection approaching its worst-case runtime, where it has to consider all test cases over all individuals. Helmuth et al. [20] find that lexicase selection often selects a single individual after considering only a small fraction of the total training cases, resulting in empirical runtimes that are much faster than the worst-case runtimes. On problems like 344_mv, however, where the selection dynamics approach the worst-case situation, lexicase selection will have much higher runtimes than we'd expect from empirical results. In contrast, DALex has a consistent and low runtime that scales predictably with the length of individuals' error vectors. The high runtimes and degraded approximiation performance for plexicase on large datasets are new findings that call for additional investigation.

4.4   Learning Classifier Systems

Finally, to show that DALex can reproduce the success of batch-lexicase, we use the learning classifier system (LCS) from Aenugu et al. [1] We choose the led24 dataset [3] for the stark contrast between the performances of lexicase selection and batch-lexicase selection. The led24 problem [3], taken from the UCI ML Repository, tests the ability of selection methods to cope with noise. This dataset, consisting of 3200 instances, contains 7 boolean attributes corresponding to whether each of 7 LED lights in a display is lit up. It has 10 categories, the digits 0 through 9, corresponding to which digit is shown on the LED display. In this dataset, each boolean attribute has a 0.1 probability having its value inverted. The evolved rule distribution for this problem is sparse, with most rules representing fewer than 5 data instances [7]. This property makes the led24 problem unsuitable for lexicase selection and motivates the development of batch-lexicase selection. For this domain, we use a particularity pressure of 20 as determined by a preliminary hyperparameter search.

As in Aenugu et al. [1], we conduct experiments in both the full data and partial data paradigms. In the full data paradigm, we use the entire led24 dataset, leaving out 30% for testing. In the partial data paradigm, we randomly downample the dataset, removing 40% of the instances. We repeat this downsample for each run, meaning the instances available for each run will be different. We then split the remaining 60% of the instances into a 40% training set and a 20% testing set.

Figure 3 displays the results of our experiments. In both domains, we find that DALex has both the highest test accuracy of the four selection methods and generates the least number of distinct rules, which suggests that it may have better generalization ability. Finally, DALex has a significantly lower runtime in both domains than lexicase and batch-lexicase selection.

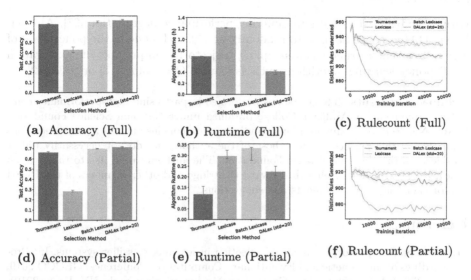

(a) Accuracy (Full)   **(b)** Runtime (Full)   **(c)** Rulecount (Full)

(d) Accuracy (Partial)   **(e)** Runtime (Partial)   **(f)** Rulecount (Partial)

Fig. 3. Performance of four selection methods on the led24 problem in the full data and partial data scenarios. Bootstrapped 95% confidence intervals are displayed as error bars for the accuracy and runtime plots and as a shaded region for the rulecount plot. DALex has approximately equal test accuracy to batch lexicase while having much lower runtime and generating the fewest rules.

5   Conclusion and Future Work

In this work, we presented the novel parent selection method DALex, which efficiently approximates the lexicase selection mechanism by using random aggregations of error vectors. The proposed method, parametrized by the particularity pressure, is flexible enough to replicate the successes of lexicase selection and its relaxed variants on multiple problem domains, and is robust to very large datasets. Furthermore, it has a consistent and low runtime, which frees up resources that can be used for the other aspects of evolution such as number of generations. We make the code used in these experiments available at https://github.com/andrewni420/DALex.

While we have used reasonable presets for the standard deviation hyperparameter in DALex, we do not anticipate that a single hyperparameter setting will work for all cases or even for all generations during the course of a GP run. Therefore, future work could determine the optimal standard deviation using hyperparameter search or even a more radical paradigm such as autoconstruction [41,43], Additionally, since DALex computes a randomly weighted fitness function, it can be combined with selection methods other than pure elitist selection, such as tournament or fitness-proportional selection [11] to increase diversity. In fact, the weighted aggregation formulation of DALex is not confined to parent selection algorithms. The concept of randomly weighted aggregation could also be used to transform the losses for each training case in deep learning. By introducing more stochasticity into training, this could enable models to be trained with large batch sizes without suffering performance degradation [27]. Alternatively, DALex could be used to randomly weight each node in the policy head of a reinforcement learner when sampling trajectories, improving the exploration of off-policy learning algorithms such as Double Q-learning [13] or SAC [12].

Acknowledgements. This work was performed in part using high-performance computing equipment at Amherst College obtained under National Science Foundation Grant No. 2117377. Any opinions, findings, and conclusions or recommendations expressed in this publication are those of the authors and do not necessarily reflect the views of the National Science Foundation. The authors would like to thank Ryan Boldi, Bill Tozier, Tom Helmuth, Edward Pantridge and other members of the PUSH lab for their insightful comments and suggestions.

References

1. Aenugu, S., Spector, L.: Lexicase selection in learning classifier systems. In: Proceedings of the Genetic and Evolutionary Computation Conference. GECCO 2019, pp. 356–364. Association for Computing Machinery, New York, NY, USA (2019). https://doi.org/10.1145/3321707.3321828
2. Boldi, R., et al.: Informed down-sampled lexicase selection: identifying productive training cases for efficient problem solving. Evol. Comput. 1–32 (2024). https://doi.org/10.48550/arXiv.2301.01488, to appear
3. Breiman, L., Friedman, J., Olshen, R., Stone, C.: LED Display Domain. UCI Mach. Learn. Repository (1988). https://doi.org/10.24432/C5FG61
4. Coppersmith, D., Winograd, S.: Matrix multiplication via arithmetic progressions. In: Proceedings of the Nineteenth Annual ACM Symposium on Theory of Computing. STOC 1987, pp. 1–6. Association for Computing Machinery, New York, NY, USA (1987). https://doi.org/10.1145/28395.28396
5. Deb, K., Agrawal, S., Pratap, A., Meyarivan, T.: A fast elitist non-dominated sorting genetic algorithm for multi-objective optimization: NSGA-II. In: Schoenauer, M., et al. (eds.) PPSN 2000. LNCS, pp. 849–858. Springer, Berlin Heidelberg, Berlin, Heidelberg (2000). https://doi.org/10.1007/3-540-45356-3_83
6. Deb, K., Goldberg, D.E.: An investigation of niche and species formation in genetic function optimization. In: Proceedings of the Third International Conference on Genetic Algorithms.p p. 42–50. Morgan Kaufmann Publishers Inc., San Francisco, CA, USA (1989)
7. Ding, L., Boldi, R., Helmuth, T., Spector, L.: Lexicase selection at scale. In: Proceedings of the Genetic and Evolutionary Computation Conference Companion. p. 2054–2062. GECCO'22, Association for Computing Machinery, New York, NY, USA (2022). https://doi.org/10.1145/3520304.3534026
8. Ding, L., Pantridge, E., Spector, L.: Probabilistic lexicase selection. In: Proceedings of the Genetic and Evolutionary Computation Conference. GECCO 2023, pp. 1073–1081. Association for Computing Machinery, New York, NY, USA (2023). https://doi.org/10.1145/3583131.3590375
9. Ding, L., Spector, L.: Optimizing neural networks with gradient lexicase selection. In: International Conference on Learning Representations (2022). https://doi.org/10.48550/arXiv.2312.12606
10. Dolson, E.: Calculating lexicase selection probabilities is NP-hard. In: Proceedings of the Genetic and Evolutionary Computation Conference. GECCO 2023. ACM (2023). https://doi.org/10.1145/3583131.3590356

11. Goldberg, D.E., Deb, K.: A comparative analysis of selection schemes used in genetic algorithms. Found. Genet. Algorithms **1**, 69–93 (1991). https://doi.org/10.1016/B978-0-08-050684-5.50008-2
12. Haarnoja, T., Zhou, A., Abbeel, P., Levine, S.: Soft actor-critic: off-policy maximum entropy deep reinforcement learning with a stochastic actor. In: Dy, J., Krause, A. (eds.) Proceedings of the 35th International Conference on Machine Learning. Proceedings of Machine Learning Research, vol. 80, pp. 1861–1870. PMLR (2018). https://proceedings.mlr.press/v80/haarnoja18b.html
13. Hasselt, H.: Double Q-learning. In: Advances in neural information processing systems. vol. 23, pp. 2613–2621 (2010)
14. He, K., Zhang, X., Ren, S., Sun, J.: Deep residual learning for image recognition. In: Proceedings of the IEEE Conference on Computer Vision and Pattern Recognition (CVPR) (2016)
15. Helmuth, T., Abdelhady, A.: Benchmarking parent selection for program synthesis by genetic programming. In: Proceedings of the 2020 Genetic and Evolutionary Computation Conference Companion. GECCO 2020, pp. 237–238. Association for Computing Machinery, New York, NY, USA (2020). https://doi.org/10.1145/3377929.3389987
16. Helmuth, T., Lengler, J., La Cava, W.: Population diversity leads to short running times of lexicase selection. In: Rudolph, G., Kononova, A.V., Aguirre, H., Kerschke, P., Ochoa, G., Tušar, T. (eds.) PPSN 2022. LNCS, pp. 485–498. Springer, Cham (2022). https://doi.org/10.1007/978-3-031-14721-0_34
17. Helmuth, T., McPhee, N.F., Spector, L.: Lexicase selection for program synthesis: a diversity analysis. In: Riolo, R., Worzel, B., Kotanchek, M., Kordon, A. (eds.) Genetic Programming Theory and Practice XIII. GEC, pp. 151–167. Springer, Cham (2016). https://doi.org/10.1007/978-3-319-34223-8_9
18. Helmuth, T., McPhee, N.F., Spector, L.: Program synthesis using uniform mutation by addition and deletion. In: Proceedings of the Genetic and Evolutionary Computation Conference. GECCO 2018, pp. 1127–1134. Association for Computing Machinery, New York, NY, USA (2018). https://doi.org/10.1145/3205455.3205603
19. Helmuth, T., Pantridge, E., Spector, L.: Lexicase selection of specialists. In: Proceedings of the Genetic and Evolutionary Computation Conference. GECCO 2019, pp. 1030–1038. Association for Computing Machinery, New York, NY, USA (2019). https://doi.org/10.1145/3321707.3321875
20. Helmuth, T., Pantridge, E., Spector, L.: On the importance of specialists for lexicase selection. Genet. Program Evolvable Mach. **21**(3), 349–373 (2020). https://doi.org/10.1007/s10710-020-09377-2
21. Helmuth, T., Spector, L.: General program synthesis benchmark suite. In: Proceedings of the 2015 Annual Conference on Genetic and Evolutionary Computation. GECCO 2015, p. 1039–1046. Association for Computing Machinery, New York, NY, USA (2015). https://doi.org/10.1145/2739480.2754769
22. Helmuth, T., Spector, L.: Explaining and exploiting the advantages of down-sampled lexicase selection. In: Proceedings of ALIFE 2020: The 2020 Conference on Artificial Life, pp. 341–349 (2020). https://doi.org/10.1162/isal_a_00334
23. Helmuth, T., Spector, L.: Problem-solving benefits of down-sampled lexicase selection. Artif. Life **27**(3–4), 183–203 (2022). https://doi.org/10.1162/artl_a_00341
24. Helmuth, T., Spector, L., Matheson, J.: Solving uncompromising problems with lexicase selection. IEEE Trans. Evol. Comput. **19**(5), 630–643 (2015). https://doi.org/10.1109/TEVC.2014.2362729

25. Hernandez, J.G., Lalejini, A., Dolson, E., Ofria, C.: Random subsampling improves performance in lexicase selection. In: Proceedings of the 2019 Genetic and Evolutionary Computation Conference Companion. GECCO 2019, pp. 2028–2031. Association for Computing Machinery, New York, NY, USA (2019). https://doi.org/10.1145/3319619.3326900

26. Holland, J.H.: Adaptation in Natural and Artificial Systems: An Introductory Analysis with Applications to Biology, Control, and Artificial Intelligence. MIT press, Cambridge (1992)

27. Keskar, N.S., Mudigere, D., Nocedal, J., Smelyanskiy, M., Tang, P.T.P.: On large-batch training for deep learning: generalization gap and sharp minima. CoRR (2016). http://arxiv.org/abs/1609.04836

28. Klein, J., Spector, L.: Genetic programming with historically assessed hardness. In: Genetic Programming Theory and Practice VI, pp. 1–14. Springer, Boston (2008). https://doi.org/10.1007/978-0-387-87623-8_5

29. Krizhevsky, A., Hinton, G., et al.: Learning multiple layers of features from tiny images. Technical report. University of Toronto (2009)

30. La Cava, W., Helmuth, T., Spector, L., Moore, J.H.: A probabilistic and multi-objective analysis of lexicase selection and ε-lexicase selection. Evol. Comput. **27**(3), 377–402 (2019). https://doi.org/10.1162/evco_a_00224

31. La Cava, W., et al.: Contemporary symbolic regression methods and their relative performance. In: Vanschoren, J., Yeung, S. (eds.) Proceedings of the Neural Information Processing Systems Track on Datasets and Benchmarks. vol. 1. Curran (2021)

32. La Cava, W., Spector, L., Danai, K.: Epsilon-lexicase selection for regression. In: Proceedings of the Genetic and Evolutionary Computation Conference 2016. GECCO 2016, pp. 741–748. Association for Computing Machinery, New York, NY, USA (2016). https://doi.org/10.1145/2908812.2908898

33. McKay, R.I.B.: Fitness sharing in genetic programming. In: Proceedings of the 2nd Annual Conference on Genetic and Evolutionary Computation. GECCO 2000, pp. 435–442. Morgan Kaufmann Publishers Inc., San Francisco, CA, USA (2000)

34. de Melo, V.V., Vargas, D.V., Banzhaf, W.: Batch tournament selection for genetic programming: the quality of lexicase, the speed of tournament. In: Proceedings of the 2019 Genetic and Evolutionary Computation Conference. GECCO 2019, pp. 994–1002. Association for Computing Machinery, New York, NY, USA (2019). https://doi.org/10.1145/3321707.3321793

35. Metevier, B., Saini, A.K., Spector, L.: Lexicase selection beyond genetic programming. In: Banzhaf, W., Spector, L., Sheneman, L. (eds.) Genetic Programming Theory and Practice XVI. GEC, pp. 123–136. Springer, Cham (2019). https://doi.org/10.1007/978-3-030-04735-1_7

36. Moore, J.M., Stanton, A.: Tiebreaks and diversity: isolating effects in lexicase selection. In: Proceedings of ALIFE 2018: The 2018 Conference on Artificial Life, pp. 590–597 (2018). https://doi.org/10.1162/isal_a_00109

37. Pantridge, E., Spector, L.: Code building genetic programming. In: Proceedings of the 2020 Genetic and Evolutionary Computation Conference. GECCO 2020, pp. 994–1002. Association for Computing Machinery, New York, NY, USA (2020). https://doi.org/10.1145/3377930.3390239

38. Romano, J.D., et al.: PMLB v1.0: an open source dataset collection for benchmarking machine learning methods (2021). https://doi.org/10.48550/arXiv.2012.00058

39. Simonyan, K., Zisserman, A.: Very deep convolutional networks for large-scale image recognition. In: Proceedings of the International Conference on Learning Representations (2015). https://doi.org/10.48550/arXiv.1409.1556
40. Sobania, D., Rothlauf, F.: Program synthesis with genetic programming: the influence of batch sizes. In: Medvet, E., Pappa, G., Xue, B. (eds.) EuroGP 2022: Genetic Programming. Lecture Notes in Computer Science, vol. 13223, pp. 118–129. Springer, Cham (2022)
41. Spector, L.: Autoconstructive evolution: Push, PushGP, and Pushpop. In: Proceedings of the 2001 Genetic and Evolutionary Computation Conference, GECCO-2001, pp. 137–146. Morgan Kaufmann Publishers, San Francisco, CA (2001)
42. Spector, L., Ding, L., Boldi, R.: Particularity. In: Winkler, S., Trujillo, L., Ofria, C., Hu, T. (eds.) Genetic Programming Theory and Practice XX. Genetic and Evolutionary Computation, Springer, Singapore (2023). https://doi.org/10.1007/978-981-99-8413-8_9 to appear
43. Spector, L., Robinson, A.: Genetic programming and autoconstructive evolution with the push programming language. Genet. Program Evolvable Mach. 3(1), 7–40 (2002). https://doi.org/10.1023/A:1014538503543
44. Stanton, A., Moore, J.M.: Lexicase selection for multi-task evolutionary robotics. Artif. Life 28(4), 479–498 (2022). https://doi.org/10.1162/artl_a_00374
45. Stephens, T.: gplearn (2023). https://github.com/trevorstephens/gplearn
46. Urbanowicz, R.J., Moore, J.H.: Learning classifier systems: a complete introduction, review, and roadmap. J. Artif. Evol. Appl. 2009, 1–25 (2009). https://doi.org/10.1155/2009/736398

Enhancing Large Language Models-Based Code Generation by Leveraging Genetic Improvement

Giovanni Pinna[ID], Damiano Ravalico[ID], Luigi Rovito[✉][ID], Luca Manzoni[ID], and Andrea De Lorenzo[ID]

University of Trieste, 34127 Trieste, TS, Italy
{giovanni.pinna,damiano.ravalico,luigi.rovito}@phd.units.it,
{lmanzoni,andrea.delorenzo}@units.it

Abstract. In recent years, the rapid advances in neural networks for Natural Language Processing (NLP) have led to the development of Large Language Models (LLMs), able to substantially improve the state-of-the-art in many NLP tasks, such as question answering and text summarization. Among them, one particularly interesting application is automatic code generation based only on the problem description. However, it has been shown that even the most effective LLMs available often fail to produce correct code. To address this issue, we propose an evolutionary-based approach using Genetic Improvement (GI) to improve the code generated by an LLM using a collection of user-provided test cases. Specifically, we employ Grammatical Evolution (GE) using a grammar that we automatically specialize—starting from a general one—for the output of the LLM. We test 25 different problems and 5 different LLMs, showing that the proposed method is able to improve in a statistically significant way the code generated by LLMs. This is a first step in showing that the combination of LLMs and evolutionary techniques can be a fruitful avenue of research.

Keywords: Evolutionary Computation · Evolutionary Algorithms · Large Language Models · Artificial Intelligence · Machine Learning · Neural Networks · Code Generation · Genetic Improvement · Grammatical Evolution · Genetic Programming

1 Introduction

Designing and developing software is a complex and demanding activity, which requires specific skills, knowledge, and expertise. Hence, tools to help the programmer in producing the code have been developed during the decades, with recent advances providing tools for automatic code generation, including LLM-based ones, like Copilot [46,51]. In general, the text generation abilities of LLMs can be leveraged to help developers in generating code starting from a description of its expected behavior. However, when the description (i.e., prompt) of the problem is poorly-formulated and not well-defined or the problem itself is non-trivial, the generated code may be incorrect or incomplete. Even in that case, the code can still provide a useful starting point to provide a complete and correct solution.

In this paper, we define an evolutionary method—Genetic Improvement (GI)—that employs a collection of user-provided test cases (as input-output pairs) starting from the code generated by LLMs having only a textual prompt. The output of the LLM is also used to influence the method employed by GI. For it, we employ Grammatical Evolution (GE) with a grammar that is dynamically built for each problem based on the code provided by the LLM. This design choice enables us to automatically implement a grammar that, with high probability, is coherent with the problem without the drawbacks of either using a very large grammar or having to manually define a problem-specific one. The proposed method has a large applicability since it can be used to enhance any LLM-generated code, with the only requirement being a set of input-output pairs for each problem. The support of languages different from Python can be done by defining the grammar of the language and the language-specific parts of specializing the grammar.

We test 25 different problems from PSB2 [16,17] with 5 LLMs and we show that the proposed method is able to improve the correctness of code generated by LLMs in a statistically significant way. The research questions we aim to answer after our experimental analysis are detailed as follows:

RQ1 Do LLMs suffice to automatically generate code that is correct regardless of the complexity of the tackled problem?
RQ2 Is it possible to improve the correctness of the code generated by LLMs by employing a GI-based technique?

The main contributions of our work are, thus, the followings: (i) we implemented a dynamic grammar definition that can specialize a grammar for the specific problem, and (ii) we introduced an evolutionary method that improves the correctness of code generated by LLMs. While the proposed method is implemented for the Python language, the principles are not tied to it and are applicable to other programming languages.

In Sect. 2 we provide a literature review of this broad research field, in Sect. 3 we delve into our proposed method, in Sect. 4 we show the outcome of our experimental analysis, in Sect. 5 we try to give an answer to our research questions, and in Sect. 6 we do a brief recap with final conclusions.

2   Related Works

2.1   Large Language Models

The development of neural network-based architectures, particularly the Transformer model [52], marked a significant turning point in Natural Language Processing (NLP). This model, relying exclusively on attention mechanisms, paved the way for advancements in text generation and comprehension. Subsequent to this, various tools and models, such as BERT [10], have been developed, all trained on extensive text corpora to enhance their ability to infer missing text from input sequences.

In 2023, Meta AI released LLaMA [49], a family of pre-trained LLMs open-sourced for research and development. These models, with parameters ranging from 7 to 65 billion, were trained on vast quantities of textual data. A notable derivative, Alpaca [48], developed by Stanford University, is a fine-tuned version of LLaMA that employs the self-instruct method for instruction tuning [54]. Alpaca, specifically the 7B and 13B versions, was trained on instruction-following demonstrations, providing capabilities akin to other chatbots.

Additionally, LLaMA 2 [50], an enhanced version with up to 70 billion parameters, was trained on 40% more data than its predecessor, further advancing the field.

The contribution of OpenAI to LLMs, particularly with GPT3 [6] and its subsequent versions, GPT3.5 and GPT4 [37], showcases significant advancements in text generation and comprehension. These models, based on decoder-only Transformer architectures, vary in the number of parameters and scale of training data. GPT4, available through ChatGPT Plus and OpenAI APIs, continues to demonstrate remarkable capabilities in various text-based tasks, although limitations remain due to its predictive nature and the unavailability of its training data and procedures for public scrutiny.

2.2   Code Generation

The research field concerning code generation was born even before the first electronic computers were released [32], and has already found great interest in the scientific community since the second half of the Twentieth Century with program synthesis [4,14,29,30] and program transformation [55].

Recently, automatic code generation tools were proposed with the purpose of implementing design patterns [7], generating efficient code based on Event-B formal specifications [33], generating C code from mathematical models [42], generating code based on Generative Adversarial Networks (GANs) [47], generating code in Verilog Hardware Description Language (VHDL) based on Unified Modeling Language (UML) specifications [35] or high-level hardware descriptions [25], and generating code for web applications based on the Model-View-Controller (MVC) pattern [39].

Automatic code generation has found great interest even in the field of Evolutionary Computation (EC). Specifically, Genetic Programming (GP) [20] is probably the most famous evolutionary algorithm that was initially developed with the purpose of evolving programs. Then, a GP method defined as Grammatical Evolution (GE) [36,43] was also introduced and gained popularity because of its ability to represent and evolve programs that are compliant with a given grammar.

GE was adopted to generate VHDL code representing digital circuits [19]. In [27], an evolutionary algorithm was implemented to generate code for Programmable Logic Controllers (PLCs) in controlling processes. In [56], an optimized hybrid evolutionary algorithm was employed to predict code running time and speed up automatic code optimization for Ansor [57], which is an auto-scheduling system that extends Apache TVM [9]. In [44], a hybrid Genetic Algorithm (GA) was leveraged to automate parallel code generation for regular control problems. Cartesian Genetic Programming (CGP) [34] was employed in [53] to evolve machine code for a simpler implementation of the MOVE processor. Multi-objective linear GP was used in [45] to generate assembly driver routines for devices belonging to micro-controller-based systems.

The recent advent of LLMs has additionally enlarged the scope of automatic code-generation techniques. Especially, LLaMA and ChatGPT, have expanded automatic code generation. ChatGPT has demonstrated proficiency in this domain but has limitations [3]. Studies on the effectiveness of LLMs include an analysis of Copilot for coding support [51], an empirical assessment of Chat-GPT generated code [24], and a study on ChatGPT-like tools non-determinism impacting scientific validity [38].

Evaluating code generation is challenging, prompting the introduction of several benchmarks. In [2], the LLMs prior to LLaMA and ChatGPT were tested against two benchmark datasets to identify their limitations. The PSB2 benchmark suite [16,17], featuring 25 updated general program synthesis problems, assesses code generation methods. For instance, the effectiveness of Copilot was compared with evolutionary approaches like GP using PSB2 [46]. The HUMANEVAL set [8], designed to measure functional correctness in code synthesis from doc-strings, tested LLMs like Codex. EvalPlus [23] further extends the test cases of HUMANEVAL to rigorously evaluate LLM-synthesized code's functional correctness.

Although numerous code generation tools, including LLMs-based ones, have demonstrated impressive capabilities, their effectiveness in generating completely correct code varies with problem complexity and, for LLMs, prompt quality. Genetic Improvement (GI) [22], a GP method, optimizes existing code through evolutionary processes. GI has been applied to enhance auto-generated code in languages like C++ [40], C, Java [31], and Python [11]. Tools like [1,5] focus on bug removal and time efficiency optimization. Recent studies have explored integrating GI methods with LLMs, evolving code snippets generated and modified by LLMs [26]. For example, [41] used recent LLMs to create new hybrid swarm intelligence optimization algorithms.

Besides evolving the generated code, also the prompt itself can be evolved and refined. This is because writing an effective prompt is not trivial and, especially for an LLM, adopting a well-written and descriptive-enough prompt could yield totally different results. EvoPrompt [15] was developed as a framework for discrete prompt optimization in which LLMs are employed to evolve prompts. Promptbreeder [12] was presented as a general-purpose self-referential self-improvement technique that opportunely evolves and adapts prompts for a specific domain.

In this work, we propose a GI method based on GE in which we improve the correctness of code generated by the most recent and performing LLMs.

3   Proposed Approach

In this section, we introduce the technique used to improve the LLM output via GI. We can identify the following main steps:

- **Encoding of the LLM output.** The output of the LLM is processed to extract the code to be improved and encode it in a genome;
- **Dynamic Grammar Definition.** Both the prompt and the initial solution generated by the LLM is used to specialize a grammar to the specific problem;
- **Evolution.** GE is used to evolve the genome using the specialized grammar.

The second and third steps are what compose GI properly, with the first step being a necessary preprocessing phase. All these steps will be detailed below.

3.1   Encoding of the LLM Output

As a first step, the textual description of the problem is given as input to an LLM in order to obtain a textual answer that contains, among possibly additional text, the code that the LLM consider to be the solution to the problem.

We recall that, even when directed by textual instructions, the output of an LLM can contain text that is extraneous with respect to the actual code needed to solve the problem at hand. Hence, the first phase is the one of extracting the code from the textual answer. To this end, we employ specialized tokens called *tags* that delimit the starting and ending position of the code in the text.

Once the code has been obtained, a series of preprocessing steps are performed in order to allow its encoding in a way that is suitable for GI:

1. Removal of comments, since they are not necessary for actually executing the code. In the current implementation, this is performed via regular expressions matching all comments.
2. Once the comments have been removed, the procedure/function actually representing the solution is extracted. In this step, we need to ensure that the code is actually syntactically correct. As a simple heuristic, we start with the full body of the function, we check if it is syntactically correct and, if not, we remove one line at a time starting from the end until a syntactically

valid solution has been found. In the worst case, this step will produce an empty function body which is syntactically correct or can be made so in an automated way (e.g., adding a `pass` for Python).

3. With a syntactically valid code it is now possible to represent the code as an Abstract Syntax Tree (AST). In this representation the name of the variables is immaterial, so their names are replaced with v_i, with i ranging from 0 the number of variables minus one. This allows for the evolution process not having to use a grammar with specialized variable names. Since the name of the function actually containing the solution to the problem is immaterial, it is replaced by a default value which, in our implementation, is `evolve`.

The additional steps to be taken depend on the specifics of the implementation and the language since the AST should be then transformed in a way that is manageable by the GI technique employed. In particular, we may impose restrictions on the grammar used to produce the AST (e.g., having it represent only a subset of the entire language) in order to help the successive GI step. In the specific implementation used in this paper the restriction on the grammar is imposed by only parsing a subset of the Python language, and the additional preprocessing step consists of replacing indents and newlines with specific tags that are then used by the GE implemented with the `ponyge2` library [11].

3.2 Dynamic Grammar Definition

Once the LLM-produced code is correctly encoded, it can be used, together with the textual description of the problem, to modify an existing (general) grammar to include problem-specific constants and words.

Let $\mathcal{G}$ be a grammar consisting of all the main constructs of the programming language used to express the solution to the problem. Notice that the grammar $\mathcal{G}$ might not be the standard formal grammar for a language, but will have some biases and changes necessary for the evolutionary process to work. First of all, $\mathcal{G}$ will be designed in such a way that, when used in a generative fashion, there is a bias towards small and actually manageable programs. This is generally obtained by giving additional derivations to expansion containing terminal instead of non-terminals, skewing a random selection toward shorter individuals, but without changing the recognized (or generated) language. In addition to that, it is possible to make the grammar generate code already including, for example, commonly used libraries and functions without having to generate "from scratch" their names.

The main idea is to add additional terminal symbols to the grammar $\mathcal{G}$ that are problem-specific. In particular, there are two main classes of symbols that can be useful to guide the search: constants and functions. Constants can be present in the problem description (e.g., a description containing "find all multiple of 37 less than 1000" has two constants) and finding them during the evolutionary process can be a difficult task while inserting them from the problem description makes easy to use in the resulting solution. Functions and libraries can help not "reinvent the wheel" by generating existing library code from scratch, but finding which libraries can be useful is a complex task that requires some knowledge of the language ecosystem (and not only of its grammar). Being trained on a large corpus of text—including code—makes the LLM a useful source of information on which libraries can be useful. Hence, the following steps are taken to extract information from the problem description and the LLM output:

- Extraction of imported libraries, function names, string and numbers from the LLM output. This ensures that libraries—and the functions they offer— that are useful to construct the problem's solution can be used in the GI step;
- Use of the problem description to extract constants (either strings or numbers);
- Use of KeyBERT [13] to extract keywords from the problem description to be used as terminals in the grammar, always in the form of strings.

Hence the problem-specific grammar $\mathcal{G}_P$ is obtained by augmenting $\mathcal{G}$ with the previously extracted symbols. In addition to that, it is possible to change the probabilities of generating the extracted symbols (making their presence in a solution more probable) by replicating the added derivation multiple times.

One clear advantage of this approach is that the solutions found by the evolutionary process can more easily use the existing library functions and there is no need to evolve "from scratch" constants that can easily be inferred by the problem description. One drawback is, however, that the quality of the grammar is strongly dependent upon the quality of the description provided and the quality of the output produced by the LLM. An incorrect but generally reasonable output (e.g., containing calls to useful library functions) can be helpful in guiding the search. A completely incorrect output which is extremely far from a real solution can make the evolutionary process worse by increasing the probability of using constants and functions that are not useful.

3.3   Evolution

Since GE is employed as the evolutionary process to enhance existing solutions, it is necessary to define two main components: the creation of the initial population and the evaluation of individuals. The other steps of the evolutionary process, specifically selection, recombination, and mutation, are not specific to the proposed method and, therefore, will not be discussed in detail here.

The initial population is initialized by the solutions generated by the LLM once encoded as detailed before. Since the number of solutions generated in this way can be smaller than the population size, individuals are replicated until the desired population size has been reached. This step is necessary since LLM generation of solutions can be expensive in terms of computational resources.

Regarding the fitness function used, we employ a series of test cases (as input-output pairs) provided by the user in its definition. Formally, given a set of n test cases $D_P = \{(x_1, y_1), (x_2, y_2), \ldots, (x_n, y_n)\}$ for the problem P and a program M, the fitness function is defined as:

$$F_{D_P}(M) = \sum_{i=1}^{n} [\text{out}(M, x_i) \neq y_i] \tag{1}$$

where $\text{out}(M, x_i)$ represents the output of M on input x_i and $[\cdot]$ is the indicator function that is 1 if the enclosed proposition is true and 0 otherwise. By minimizing F_{D_P} we reduce the number of incorrect outputs across the set D_P of test cases.

4 Experimental Phase

In this section, we detail our experimental phase, in which we test the following LLMs: Alpaca 7B (A7), Alpaca 13B (A13), LLaMA 2 13B (L2), ChatGPT (CG), and GPT4 (G4).

4.1 Language and Grammar

While the proposed method has been described in a general way, in the experimental phase we focused on the Python language. The choice was mainly due to the fact that popular languages provide more training material to LLM, thus increasing their ability to generate meaningful solutions and the availability of libraries for parsing and analysis of the code.

The grammar we employed is actually a subset of the entire Python grammar whose structure is shortly defined here. First of all, the grammar always generates a series of imports followed by a call to the `evolve` function, which receives a certain number of parameters as input and contains some code as the body. The code can include one or more statements, which are divided into conditional statements (if, for, while) and non-conditional statements. Especially, a non-conditional statement can be either a return, an assignment, a print, a break keyword, a continue keyword, a pass keyword, a raise, or an instance method called on an object. A non-terminal representing a value can be either a problem parameter, a string, a number, a Boolean, a null, a logical or mathematical combination of two variables, a list, a tuple, a list comprehension, a tuple comprehension, or a function called on variables. The complete base grammar in Backus-Naur Form (BNF) for the subset of Python that we use can be found in our repository (see Sect. 4).

4.2 Problems

In order to evaluate our approach, we chose to utilize a widely-used problem dataset, PSB2 [16,17]. This dataset offers a comprehensive set of problems (Table 1), serving as a benchmark for validating automatic code generation algorithms. We performed an initial evaluation of the complexity of the coding problems from PSB2. This evaluation proceeds as follows: initially, for each problem, we prompt the LLM to generate a solution. This step is independently repeated 10 times with the same prompt. If the LLM correctly solves the problem (i.e., the generated code is completely correct with respect to the examples given by the user) we do not consider it for further improvement. Such problems are denoted by ✓ in Table 2.

Table 1. List of PSB2 problems.

Name	Alias	Name	Alias
Basement	BS	Bouncing Balls	BB
Bowling	BW	Camel Case	CC
Coin Sums	CS	Cut Vector	CV
Dice Game	DG	Find Pair	FP
Fizz Buzz	FB	Fuel Cost	FC
Greatest Common Divisor	GD	Indices of Substring	IS
Leaders	LD	Luhn	LH
Mastermind	MM	Middle Character	MC
Paired Digits	PD	Shopping List	SL
Snow Day	SD	Solve Boolean	SB
Spin Words	SW	Square Digits	SQ
Substitution Cipher	SC	Twitter	TW
Vector Distance	VD		

In our context, the problem description is crucial since it is the main component of the prompt for the LLMs. Since multiple problem descriptions might be available[1], among them we adopt the official one available in the PSB2 paper itself [16,17].

4.3 Experimental Settings and Results

We request the LLM to solve each problem 10 times, producing 10 individuals that are then replicated 100 times each to meet the required population size. The fitness is computed by adopting 1000 test cases as training data and 1000 test cases as test data. For each combination of problem and LLM, we conduct 10 separate GI runs starting from the same initial population. In all our experiments, the test cases are obtained via the psb2 library [16,17]. Regarding the parameters used for the evolution, we use 1000 as population size, which is

---

[1] The PSB2 paper also includes an external hyperlink to a file that contains the same problems but with different descriptions (e.g., FB description in the external table details that the output is printed, while the description in the paper itself details that the output is returned, which is more coherent with the original purpose of PSB2).

evolved for 100 generations. We set an initial maximum depth of 15 and a maximum depth of 30 for the AST.[2] We do not impose a wrap-around limit for the genomes and we set a maximum codon size of 200. The selection is performed using tournament selection with pressure 50, ensuring a population that remains syntactically close to the solution obtained from the LLM. Both crossover and mutation are sub-tree operators, applied with a probability of 0.80.

In order to test that GI is better than the self-correction ability of LLMs we implemented the following procedure, as described in [26]. For each LLM, we request it to solve a given problem (a method denoted as $\mathcal{L}$) and then we evaluate the correctness of the output code. When the code is incorrect we ask the LLM to self-correct its output (a method denoted as $\mathcal{L}^+$). During a self-correction run, we request the LLM to incrementally correct the output code 10 times. In each iteration, we supply 20 pairs of input and output examples that were not resolved in the last iteration. To provide these examples, we instruct the LLM with the prefix `Make sure that`: followed by the examples formatted as {`input`} -> {`output`}. We perform 2 self-correction runs and we store the maximum value among the median values retrieved from those runs.[3] In this way, we select the best result obtained by using self-correction of the LLM without employing GI. Our Python code is publicly available online.[4]

In Table 2 we show the median number (scaled in $[0, 1]$) of passed test cases for $\mathcal{L}$ and $\mathcal{L}^+$. Statistical significance is assessed by using a Wilcoxon-Mann-Whitney test [28] with $\alpha = 0.05$. For more than two comparisons a preliminary Kruskal-Wallis test [21] with $\alpha = 0.05$ is performed and then, if there is a statistically significant difference, a Holm-Bonferroni correction [18] with $\alpha = 0.05$ is employed. For $\mathcal{L}$ and GI, the sample size is 10. For $\mathcal{L}^+$, the sample size is 2.

For each LLM, we highlight, in the table, both the problems that are directly solved (✓) and the ones that are not parsed by the grammar (np), which is due to the fact that the employed grammar does not capture the entirety of the Python language. In principle, we could parse all the problems by defining an exhaustive grammar. Defining an exhaustive grammar leads to a larger search space that could slow down the evolutionary processes. However, the employed grammar defines a large enough subset of the Python language, giving a reasonable trad-off between expressiveness and complexity of the search space.

For A7 and A13, both $\mathcal{L}$ and $\mathcal{L}^+$ lead to incorrect code and GI provides a significantly improved version of it for all the parsed problems. L2, on average, performs slightly better than A7 and A13, and GI provides a significantly improved version of the code for almost all the parsed problems. As expected, CG and G4 directly solve more problems than the smaller LLaMA-based models. Even in this case, GI obtains a significant improvement for almost all the parsed problems that are not directly solved by them. As regards A7, there is a high

---

[2] The BW problem has an initial maximum depth of 25 and maximum depth of 40 since the initial solution requires a depth greater than 15.

[3] The number of repetitions is constrained by the limitations of our available budget.

[4] https://github.com/dravalico/LLMGIpy.

Table 2. Median number (scaled in $[0,1]$) of passed test cases for each P and LLM. We highlight in bold the methods that are statistically significant better for the same problem and LLM. Problems that are directly solved from $\mathcal{L}$ are indicated with ✓. Problems that are not parsed for the GI are indicated with np (not parsed). With a 0 we denote problems that do not pass any test cases.

Problem	A7			A13			L2			CG			G4		
	$\mathcal{L}$	$\mathcal{L}^+$	GI	$\mathcal{L}$	$\mathcal{L}^+$	GI	$\mathcal{L}$	$\mathcal{L}^+$	GI	$\mathcal{L}$	$\mathcal{L}^+$	GI	$\mathcal{L}$	$\mathcal{L}^+$	GI
BS	0.01	0	**0.58**	0	0	**0.20**	0.13	0.00	**0.20**	✓	—	—	✓	—	—
BB	0	0	np	0	0	np	0	0	np	0.00	0.00	**0.06**	0.00	0.04	**0.06**
BW	0	0	np	0	0	**0.03**	0	0	**0.06**	0.00	0.00	**0.03**	0.04	0.00	0.04
CC	0	0	**0.31**	0	0	np	0	0	**0.31**	✓	—	—	✓	—	—
CS	0	0	np	0	0	0	0	0	0	0	0	0	0	0	0
CV	0	0	np	0	0	0	0	0	np	0	0	0	0	0	0
DG	0	0	np	0	0	np	0	0.00	0.00	0.07	0.00	**0.17**	0.15	0.00	**0.17**
FP	0	0	np	0	0	np	0	0	0	0	0	0	0	0	0
FB	0	0	np	0.94	0	**0.94**	✓	—	—	✓	—	—	✓	—	—
FC	0	0	np	0	0	np	0.18	1.00	**1.00**	✓	—	—	✓	—	—
GD	0	0	np	0	0	np	0	0	**0.59**	✓	—	—	✓	—	—
IS	0	0	np	0.16	0	**0.67**	✓	—	—	✓	—	—	✓	—	—
LD	0	0	**0.09**	0.05	0	**0.09**	0	0	**0.09**	✓	—	—	✓	—	—
LH	0	0	**0.04**	0	0	**0.06**	0	0	**0.04**	0.52	0.00	**1.00**	0.00	0	0.08
MM	0	0	np	0	0	0	0	0	np	0	0	0	0	0	0
MC	0.50	0	np	0	0	**0.02**	✓	—	—	✓	—	—	✓	—	—
PD	0	0	**0.13**	0	0	**0.13**	0	0.01	**0.19**	✓	—	—	✓·	—	—
SL	0	0	**0.00**	0	0	**0.00**	0.00	0	**0.00**	0.00	0	**0.01**	0.00	0	**0.01**
SD	0	0	np	0	0	np	0.05	0	np	0.05	0.00	**0.06**	0.04	0.00	**0.09**
SB	0.50	0	**0.78**	0	0	np	0.49	0.00	**0.55**	0.00	0.42	np	0	0.67	0.50
SW	0	0	**0.34**	0	0	**0.34**	0.56	0.00	0.56	✓	—	—	✓	—	—
SQ	0	0	**0.01**	0	0	**0.01**	0.00	0	**0.01**	✓	—	—	✓	—	—
SC	0	0	np	0	0	**0.05**	0.04	0	**0.05**	✓	—	—	✓	—	—
TW	0.30	0	np	0.30	0	np	0.99	0	np	✓	—	—	✓	—	—
VD	0	0	np	0	0	np	0	0	np	0.96	0.00	**1.00**	**0.95**	0.00	0.95

percentage of problems that are not parsed, due to the low quality of the solutions provided by the LLM, which are not even parsable by the used grammar. Differently from other problems, TW is directly solved by OpenAI models, but the code produced by the others LLMs is not parsable by the used grammar, hence this is the only case where GI cannot provide any improvement in all models. In general, the self-correction procedure seems to be poorly effective for all the models, except for L2 on FC and both CG and G4 on SB, where in the very latter case no statistical significance is recorded. Except for G4 on VD where the LLM alone performs better, we can safely state that our GI method is never worse than both the LLM and the self-correction procedure. The general behavior is that GI is able to improve the code generated by the LLMs. However, the improvement leads to an optimal fitness value only in three cases, probably due to the size and complexity of the search space.

In Fig. 1, we show the fitness trend of GI. It is possible to observe two different behaviors, depending on the LLM used to generate the initial solutions. In particular, for smaller models (i.e., the LLaMA-based ones) there is an improvement in the fitness with the number of generations, possibly due to the fact that the initial solution is incorrect but not in a way that requires a complete rewrite—i.e., only small edits are sufficient to gain correctness. On the other hand, larger models (i.e., the OpenAI-based ones) exhibit a bimodal behavior: either the initial solution is completely correct or irremediably wrong—i.e., the number of passed test cases is close to zero and improving the code requires substantial rewriting. In these cases, we record a slight improvement (except for LH and VD where GI from `CG` is able to reach an optimum).

5   Discussion

Finally, we try to provide a brief summary of the main findings and give proper answers to our research questions:

RQ1 Do LLMs suffice to automatically generate code that is correct regardless of the complexity of the tackled problem?

The most performing LLMs (i.e., the larger models, in our case the OpenAI ones) tend to generate better code than the smaller ones, which in our investigation are the LLaMA-based ones. However, even these models often produce code that is not correct for many of the investigated problems. When incorrect code is generated, we attempt to prompt the LLM again with a small set of failed test cases, asking to modify the program so that it passes these specific tests. Despite these efforts, our findings indicate that if the model initially generates incorrect code, it is unlikely to correct itself effectively through mere re-prompting. This outcome highlights the challenges faced by LLMs when generating accurate code for a specified problem, even when examples are provided. It suggests that integrating additional techniques may be necessary to achieve the desired level of correctness.

RQ2 Is it possible to improve the correctness of the code generated by LLMs by employing a GI-based technique?

For almost all the tested problems, GI is capable of providing an improvement of the code generated by LLMs. In fact, the resulting code is generally better than simply re-prompting the LLMs, showing that GI can, in fact, provide an advantage with respect to a full LLM-based solution. The reason for these performances can be identified in the use of a specialized grammar. Such a grammar can guide the search process of GE in specific areas of the search space where better solutions can potentially be found. This mechanism, however, is strongly dependent on the quality of the initial solution provided by the LLM. While related to the actual correctness of the solution, the quality needed to specialize the grammar is the ability for the LLM to provide a reasonable set of functions, libraries, and constants to be used by GE. This is generally an easier task for an LLM with respect to the production of correct code, thus making even smaller LLMs a good source of initial solutions for the GI-based step. The effect

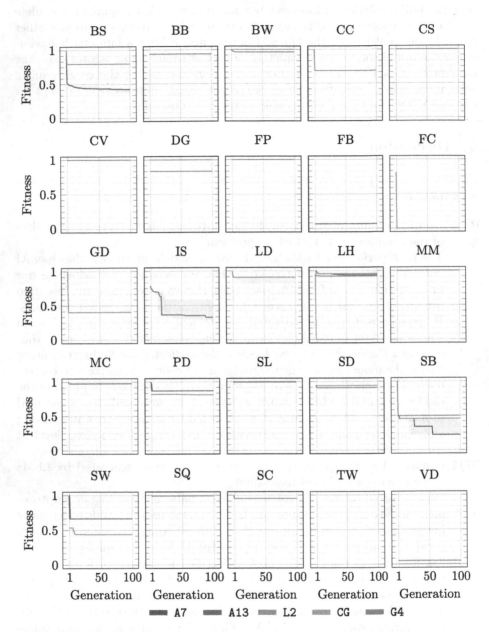

Fig. 1. Trend of the fitness (scaled in $[0, 1]$) across the generations of GI. Each line represents the median fitness across all runs. Shaded area represents the inter-quartile range. We omit the corresponding lines where no GI is performed.

is that the evolution is more effective for smaller models and simple problems (which are usually directly solved by larger models) and less evident for complex ones. A potential limitation could be the large number of test cases required to drive the evolution. Future works could focus on discovering the minimum number of test cases needed to achieve a meaningful improvement.

6   Conclusion

In this paper, we propose an evolutionary approach designed to improve the accuracy of code generated by state-of-the-art LLMs. For our approach we require the user to provide both a textual description of the problem and a set of test cases as input-output pairs. If the LLM fails to generate a solution that yields the correct outputs, we employ GI techniques to enhance the accuracy of the provided solution. To reduce the search space and enhance the effectiveness of the evolutionary process with GI, we have chosen to automatically specialize a grammar using the LLM output in order to tailor it to the specific problem. While the approach is generically applicable, in the experimental phase we focused on the Python language. For evaluating our approach, we conducted a comprehensive experimental campaign, involving 5 different LLMs and comparing GI with the self-correction ability of the LLMs. Our experimental analysis carried out on 25 problems sourced from the PSB2 dataset, showing the effectiveness of our method in enhancing the accuracy of Python code generated by the most advanced LLMs currently available.

As future research directions, we can study the effect of different specialization techniques, possibly also *reducing* the grammar according to the LLM outputs. This would allow to both use a large grammar for the initial parsing, thus increasing the number of correctly parsed solutions, and reducing the size of the search space for the GI phase. Finally, an analysis on the number of test cases needed to instruct these code correction methods would be beneficial to highlight the actual effort required by the user.

References

1. An, G., Blot, A., Petke, J., Yoo, S.: PyGGI 2.0: language independent genetic improvement framework. In: Proceedings of the 2019 27th ACM Joint Meeting on European Software Engineering Conference and Symposium on the Foundations of Software Engineering, pp. 1100–1104 (2019)
2. Austin, J., et al.: Program synthesis with large language models. arXiv preprint arXiv:2108.07732 (2021)
3. Bahrini, A., et al.: ChatGPT: applications, opportunities, and threats. In: 2023 Systems and Information Engineering Design Symposium (SIEDS), pp. 274–279 (2023)
4. Bibel, W.: Syntax-directed, semantics-supported program synthesis. Artif. Intell. **14**(3), 243–261 (1980)

5. Blot, A., Petke, J.: MAGPIE: machine automated general performance improvement via evolution of software. arXiv preprint arXiv:2208.02811 (2022)
6. Brown, T.B., et al.: Language models are few-shot learners. arXiv preprint arXiv:2005.14165 (2020)
7. Budinsky, F.J., Finnie, M.A., Vlissides, J.M., Yu, P.S.: Automatic code generation from design patterns. IBM Syst. J. **35**(2), 151–171 (1996)
8. Chen, M., et al.: Evaluating large language models trained on code. arXiv preprint arXiv:2107.03374 (2021)
9. Chen, T., et al.: {TVM}: an automated {End-to-End} optimizing compiler for deep learning. In: 13th USENIX Symposium on Operating Systems Design and Implementation (OSDI 2018), pp. 578–594 (2018)
10. Devlin, J., Chang, M.W., Lee, K., Toutanova, K.: BERT: pre-training of deep bidirectional transformers for language understanding. arXiv preprint arXiv:1810.04805 (2019)
11. Fenton, M., McDermott, J., Fagan, D., Forstenlechner, S., Hemberg, E., O'Neill, M.: PonyGE2: grammatical evolution in Python. In: Proceedings of the Genetic and Evolutionary Computation Conference Companion, pp. 1194–1201 (2017)
12. Fernando, C., Banarse, D., Michalewski, H., Osindero, S., Rocktäschel, T.: Promptbreeder: self-referential self-improvement via prompt evolution. arXiv preprint arXiv:2309.16797 (2023)
13. Grootendorst, M.: KeyBERT: minimal keyword extraction with BERT. Zenodo (2020)
14. Gulwani, S., Polozov, O., Singh, R., et al.: Program synthesis. Found. Trends® Program. Lang. **4**(1–2), 1–119 (2017)
15. Guo, Q., et al.: Connecting large language models with evolutionary algorithms yields powerful prompt optimizers. arXiv preprint arXiv:2309.08532 (2023)
16. Helmuth, T., Kelly, P.: PSB2: the second program synthesis benchmark suite. In: Proceedings of the Genetic and Evolutionary Computation Conference, pp. 785–794 (2021)
17. Helmuth, T., Kelly, P.: Applying genetic programming to PSB2: the next generation program synthesis benchmark suite. Genet. Program Evolvable Mach. **23**(3), 375–404 (2022)
18. Holm, S.: A simple sequentially rejective multiple test procedure. Scand. J. Stat. 65–70 (1979)
19. Karpuzcu, U.R.: Automatic Verilog code generation through grammatical evolution. In: Proceedings of the 7th Annual Workshop on Genetic and Evolutionary Computation, pp. 394–397 (2005)
20. Koza, J.R.: Genetic programming as a means for programming computers by natural selection. Stat. Comput. **4**, 87–112 (1994)
21. Kruskal, W.H., Wallis, W.A.: Use of ranks in one-criterion variance analysis. J. Am. Stat. Assoc. **47**(260), 583–621 (1952)
22. Langdon, W.B.: Genetic improvement of programs. In: 2014 16th International Symposium on Symbolic and Numeric Algorithms for Scientific Computing, pp. 14–19. IEEE (2014)
23. Liu, J., Xia, C.S., Wang, Y., Zhang, L.: Is your code generated by ChatGPT really correct? rigorous evaluation of large language models for code generation. arXiv preprint arXiv:2305.01210 (2023)
24. Liu, Z., Tang, Y., Luo, X., Zhou, Y., Zhang, L.F.: No need to lift a finger anymore? Assessing the quality of code generation by ChatGPT. arXiv preprint arXiv:2308.04838 (2023)

25. Liu, Z., Dou, Y., Jiang, J., Xu, J.: Automatic code generation of convolutional neural networks in FPGA implementation. In: 2016 International Conference on Field-Programmable Technology (FPT), pp. 61–68. IEEE (2016)
26. Liventsev, V., Grishina, A., Härmä, A., Moonen, L.: Fully autonomous programming with large language models. In: Proceedings of the Genetic and Evolutionary Computation Conference, GECCO 2023, pp. 1146–1155. Association for Computing Machinery, New York (2023)
27. Löppenberg, M., Schwung, A.: Self optimisation and automatic code generation by evolutionary algorithms in PLC based controlling processes. arXiv preprint arXiv:2304.05638 (2023)
28. Mann, H.B., Whitney, D.R.: On a test of whether one of two random variables is stochastically larger than the other. Ann. Math. Stat. 18(1), 50–60 (1947)
29. Manna, Z., Waldinger, R.: Knowledge and reasoning in program synthesis. Artif. Intell. 6(2), 175–208 (1975)
30. Manna, Z., Waldinger, R.J.: Toward automatic program synthesis. Commun. ACM 14(3), 151–165 (1971)
31. Marino, F., Squillero, G., Tonda, A.: A general-purpose framework for genetic improvement. In: Handl, J., Hart, E., Lewis, P.R., López-Ibáñez, M., Ochoa, G., Paechter, B. (eds.) PPSN 2016. LNCS, vol. 9921, pp. 345–352. Springer, Cham (2016). https://doi.org/10.1007/978-3-319-45823-6_32
32. Menabrea, L.F.: Sketch of the analytical engine invented by Charles Babbage, ESQ. In: Ada's Legacy: Cultures of Computing from the Victorian to the Digital Age (1843)
33. Méry, D., Singh, N.K.: Automatic code generation from Event-B models. In: Proceedings of the 2nd Symposium on Information and Communication Technology, pp. 179–188 (2011)
34. Miller, J.F., Harding, S.L.: Cartesian genetic programming. In: Proceedings of the 10th Annual Conference Companion on Genetic and Evolutionary Computation, pp. 2701–2726 (2008)
35. Moreira, T.G., Wehrmeister, M.A., Pereira, C.E., Petin, J.F., Levrat, E.: Automatic code generation for embedded systems: from UML specifications to VHDL code. In: 2010 8th IEEE International Conference on Industrial Informatics, pp. 1085–1090. IEEE (2010)
36. O'Neill, M., Ryan, C.: Grammatical evolution. IEEE Trans. Evol. Comput. 5(4), 349–358 (2001)
37. OpenAI: GPT-4 technical report. arXiv preprint arXiv:2303.08774 (2023)
38. Ouyang, S., Zhang, J.M., Harman, M., Wang, M.: LLM is like a box of chocolates: the non-determinism of ChatGPT in code generation. arXiv preprint arXiv:2308.02828 (2023)
39. Paolone, G., Marinelli, M., Paesani, R., Di Felice, P.: Automatic code generation of MVC web applications. Computers 9(3), 56 (2020)
40. Petke, J., Harman, M., Langdon, W.B., Weimer, W.: Using genetic improvement and code transplants to specialise a C++ program to a problem class. In: Nicolau, M., et al. (eds.) EuroGP 2014. LNCS, vol. 8599, pp. 137–149. Springer, Heidelberg (2014). https://doi.org/10.1007/978-3-662-44303-3_12
41. Pluhacek, M., Kazikova, A., Kadavy, T., Viktorin, A., Senkerik, R.: Leveraging large language models for the generation of novel metaheuristic optimization algorithms. In: Proceedings of the Companion Conference on Genetic and Evolutionary Computation, pp. 1812–1820 (2023)

42. Rugina, A.E., Thomas, D., Olive, X., Veran, G.: Gene-auto: automatic software code generation for real-time embedded systems. DASIA 2008-Data Syst. Aerosp. **665**, 28 (2008)

43. Ryan, C., Collins, J.J., Neill, M.O.: Grammatical evolution: evolving programs for an arbitrary language. In: Banzhaf, W., Poli, R., Schoenauer, M., Fogarty, T.C. (eds.) EuroGP 1998. LNCS, vol. 1391, pp. 83–96. Springer, Heidelberg (1998). https://doi.org/10.1007/BFb0055930

44. Sandnes, F.E., Megson, G.M.: A hybrid genetic algorithm applied to automatic parallel controller code generation. In: Proceedings of the Eighth Euromicro Workshop on Real-Time Systems, pp. 70–75. IEEE (1996)

45. Serruto, W.F., Casas, L.A.: Automatic code generation for microcontroller-based system using multi-objective linear genetic programming. In: 2017 International Conference on Computational Science and Computational Intelligence (CSCI), pp. 279–285. IEEE (2017)

46. Sobania, D., Briesch, M., Rothlauf, F.: Choose your programming copilot: a comparison of the program synthesis performance of github copilot and genetic programming. In: Proceedings of the Genetic and Evolutionary Computation Conference, pp. 1019–1027 (2022)

47. Sun, H., Nie, Y., Li, X., Huang, M., Tian, J., Kong, W.: An automatic code generation method based on sequence generative adversarial network. In: 2022 7th IEEE International Conference on Data Science in Cyberspace (DSC), pp. 383–390. IEEE (2022)

48. Taori, R., et al.: Alpaca: a strong, replicable instruction-following model. Stanford Center for Research on Foundation Models **3**(6), 7 (2023). https://crfm.stanford.edu/2023/03/13/alpaca.html

49. Touvron, H., et al.: LLaMA: open and efficient foundation language models. arXiv preprint arXiv:2302.13971 (2023)

50. Touvron, H., et al.: LLaMA 2: open foundation and fine-tuned chat models. arXiv preprint arXiv:2307.09288 (2023)

51. Vaithilingam, P., Zhang, T., Glassman, E.L.: Expectation vs experience: evaluating the usability of code generation tools powered by large language models. In: Extended Abstracts of the 2022 CHI Conference on Human Factors in Computing Systems, CHI EA 2022. Association for Computing Machinery, New York (2022)

52. Vaswani, A., et al.: Attention is all you need. In: Advances in Neural Information Processing Systems, vol. 30 (2017)

53. Walker, J.A., Liu, Y., Tempesti, G., Tyrrell, A.M.: Automatic code generation on a MOVE processor using cartesian genetic programming. In: Tempesti, G., Tyrrell, A.M., Miller, J.F. (eds.) ICES 2010. LNCS, vol. 6274, pp. 238–249. Springer, Heidelberg (2010). https://doi.org/10.1007/978-3-642-15323-5_21

54. Wang, Y., et al.: Self-instruct: aligning language models with self-generated instructions. arXiv preprint arXiv:2212.10560 (2023)

55. Ward, M.: Proving program refinements and transformations. Ph.D. thesis, University of Oxford (1989)

56. Zhang, Y., Li, Y., Wang, X.: An optimized hybrid evolutionary algorithm for accelerating automatic code optimization. In: Third International Seminar on Artificial Intelligence, Networking, and Information Technology (AINIT 2022), vol. 12587, pp. 488–496. SPIE (2023)

57. Zheng, L., et al.: Ansor: generating {High-Performance} tensor programs for deep learning. In: 14th USENIX Symposium on Operating Systems Design and Implementation (OSDI 2020), pp. 863–879 (2020)

SLIM_GSGP: The Non-bloating Geometric Semantic Genetic Programming

Leonardo Vanneschi(✉)

NOVA Information Management School (NOVA IMS), Universidade Nova de Lisboa,
Campus de Campolide, 1070-312 Lisboa, Portugal
lvanneschi@novaims.unl.pt

Abstract. Geometric semantic genetic programming (GSGP) is a successful variant of genetic programming (GP), able to induce a unimodal error surface for all supervised learning problems. However, a limitation of GSGP is its tendency to generate offspring larger than their parents, resulting in continually growing program sizes. This leads to the creation of models that are often too complex for human comprehension. This paper presents a novel GSGP variant, the Semantic Learning algorithm with Inflate and deflate Mutations (SLIM_GSGP). SLIM_GSGP retains the essential theoretical characteristics of traditional GSGP, including the induction of a unimodal error surface and introduces a novel geometric semantic mutation, the deflate mutation, which generates smaller offspring than its parents. The study introduces four SLIM_GSGP variants and presents experimental results demonstrating that, across six symbolic regression test problems, SLIM_GSGP consistently evolves models with equal or superior performance on unseen data compared to traditional GSGP and standard GP. These SLIM_GSGP models are significantly smaller than those produced by traditional GSGP and are either smaller or of comparable size to standard GP models. Notably, the compactness of SLIM_GSGP models allows for human interpretation.

Keywords: Genetic Programming · Geometric Semantic Genetic Programming · Inflate and Deflate Mutations · Model Interpretability

1 Introduction

Geometric Semantic Genetic Programming (GSGP) [19] is a variant of Genetic Programming (GP) [15] that employs Geometric Semantic Operators (GSOs) in place of the traditional crossover and mutation. Although acting on the syntax of the evolving programs, GSOs have known effects on their semantics, and induce a unimodal error surface on any supervised learning problem [28,33]. GSGP offers remarkable optimization capabilities on training data and exhibits the

Supported by SPECIES Society.

M. Giacobini et al. (Eds.): EuroGP 2024, LNCS 14631, pp. 125–141, 2024.
https://doi.org/10.1007/978-3-031-56957-9_8

ability to mitigate overfitting, as evidenced by its numerous successes in various applications over the last decade [6,9,29,30,34]. However, a well-acknowledged drawback of GSGP lies in the fact that GSOs consistently generate offspring larger than their parents. This characteristic results in a steady growth of program size throughout a GSGP run, culminating in the creation of models that are too complex for human interpretation. This paper addresses this research gap by introducing a novel GSGP variant, which, while preserving the essential the geometric properties of conventional GSGP, such as inducing a unimodal error surface, evolves significantly smaller programs. This novel variant is named the Semantic Learning algorithm with Inflate and deflate Mutations (SLIM_GSGP). The name stems from its distinctive mutation operators: in addition to an inflate mutation, which generates larger offspring akin to that of traditional GSGP, SLIM_GSGP incorporates a novel mutation operator, the deflate mutation, that produces smaller offspring than their parents. In addition to the innovative introduction of a geometric semantic mutation capable of producing offspring smaller than their parents, this research offers two additional significant contributions. It explores two distinct methods for generating random expressions whose output is constrained in a previously defined interval and two different approaches for defining a ball mutation on the semantic space. By integrating these techniques, we introduce four distinct SLIM_GSGP variants, which will be systematically compared to traditional GSGP and standard GP (stdGP).

The manuscript is organized as follows: Sect. 2 revises some previously published attempts of limiting the size of the programs evolved by GSGP and it briefly recalls the functioning of GSGP and its main properties. Section 3 presents the general idea of SLIM_GSGP and defines its four studied variants. Section 4 introduces the six real-life symbolic regression problems used as test cases and describes the employed experimental settings. Section 5 presents and discusses the obtained experimental results. Finally, Sect. 6 concludes the paper and proposes ideas for future research.

2   Literature and Methodology Overview

Literature Review. The issue of the swift expansion of code of the GSGP's evolving individuals has garnered considerable research interest in recent times. Addressing the issue has predominantly revolved around two main concepts: code simplification [16,19], and an implementation strategy that, although not minimizing code size, still facilitates the fast execution of GSGP by efficiently memorizing the evolving individuals [4,18,31]. Additional methods included evolving individuals as combinations of generic basis functions [20] and implementing an elitist version of GSGP in which offspring are incorporated into the new population only if they exhibit superior fitness compared to their parents [8]. Furthermore, competent operators were proposed as a means to decrease the required number of generations, consequently yielding a discernible advantageous impact on the size of the evolved individuals [23]. Additionally, the combination of GSGP with local search [7], and with gradient descent [24] has been demonstrated to generate smaller individuals in comparison to conventional GSGP.

Geometric Semantic GP. As in [19], we use the term *semantics* to indicate the vector of the output values of a GP individual over all the training observations. As mentioned earlier, GSGP is a variant of GP where Genetic Semantic Operators (GSOs) take the place of the conventional crossover and mutation. GSOs introduce alterations to the syntax of GP individuals, aiming for a controlled and quantifiable effect on their semantics. In the last decade, numerous contributions have clearly indicated that mutation is a more powerful operator than crossover for GSGP and a version of GSGP that uses only mutation typically returns results that are comparable, or even better, than the ones obtained when a mixture of crossover and mutation is used [5,34]. For this reason, only Geometric Semantic Mutation (GSM) is considered here. According to the definition of GSM[1] outlined in [19], given a parent function $T : \mathbb{R}^n \to \mathbb{R}$, GSM with mutation step ms returns the function $GSM(T) = T + ms \cdot (T_{R1} - T_{R2})$, where T_{R1} and T_{R2} are random functions. The motivation for considering the difference between two random functions, T_{R1} and T_{R2}, is that GSM should introduce a random perturbation with distribution centered at zero. While this was not originally defined in GSM, later studies [13,28,34] have demonstrated that limiting the codomain of T_{R1} and T_{R2} to a predefined interval such as $[0, 1]$, for instance by applying a logistic function or other types of functions to the their outputs [13], can enhance generalization ability. In this way, the application of GSM to a given parent program produces a random perturbation of its semantics; such a perturbation follows a zero-centered distribution in a user-controlled range $[-ms, ms]$, where ms is a hyperparameter. In other words, GSM corresponds to a ball mutation in the semantic space [19], and induces a unimodal error surface for any supervised learning problem (for a detailed explanation, the reader is referred to [19,28,33]).

3   Semantic Learning with Inflate and Deflate Mutations

Given that $GSM(T) = T + ms \cdot (T_{R1} - T_{R2}) = T - ms \cdot (T_{R2} - T_{R1})$ and given that T_{R1} and T_{R2} can be thought as independent random variables with the same probability distribution (over the space of expressions), T_{R1} and T_{R2} are interchangeable. In other words, if the definition of GSM was $GSM(T) = T - ms \cdot (T_{R1} - T_{R2})$, that definition would be equivalent to the one given in the previous section. Now, let us consider a chain of, for instance, three mutation events starting from an individual T. The resulting individual, $GSM(GSM(GSM(T)))$, would have a shape such as:

$$T + ms \cdot (T_{R1} - T_{R2}) + ms \cdot (T_{R3} - T_{R4}) + ms \cdot (T_{R5} - T_{R6}) \tag{1}$$

If, at the next step, we use subtraction instead of addition, we obtain an individual like:

---

[1] Note that this paper exclusively presents the definition of GSM in the context of symbolic regression problems; for GSO definitions in other domains, the reader is referred to [19].

$T + ms \cdot (T_{R1} - T_{R2}) + ms \cdot (T_{R3} - T_{R4}) + ms \cdot (T_{R5} - T_{R6}) - ms \cdot (T_{R7} - T_{R8})$.
Clearly, the genotype of this individual is larger than the one in Eq. (1). In other
words, using subtraction instead of addition allowed us to apply a mutation that
has exactly the same geometric properties as the GSM defined in Sect. 2, but
was not helpful in reducing the size of the generated individual. But what if
instead of subtracting a *new* term, we had subtracted a term that had been previously added? For instance, what if the term that we subtract to the individual
in Eq. (1) is, say, $ms \cdot (T_{R3} - T_{R4})$? In the end, the idea of reusing random trees
instead of generating new ones every time they are needed is not new (it has
already been employed, for instance, in [31]). In that case, that term could be
simplified, generating an individual with a smaller size:

$$T + ms \cdot (T_{R1} - T_{R2}) \underbrace{+ ms \cdot (T_{R3} - T_{R4})} + ms \cdot (T_{R5} - T_{R6}) \underbrace{- ms \cdot (T_{R3} - T_{R4})}$$
$$= T + ms \cdot (T_{R1} - T_{R2}) + ms \cdot (T_{R5} - T_{R6}) \qquad (2)$$

After simplification, obviously, the individual in Eq. (2) has a smaller genotype
than the one in Eq. (1). However, it is important to reinforce two concepts.
First, subtracting a term that had previously been added is *not a backtracking*
operation. In other words, it generally does not generate an individual that had
already been in the population in the previous iterations. This can be seen by
observing the sequence of mutation events that have leaded to the generation of
the individual in Eq. (1). Notice that the individual in Eq. (2) had never been
generated before in that sequence (this is general, unless the subtracted term is
the last one that had been added). Second, subtracting a term that had previously been added has the same type of effect as GSM: a perturbation of each
semantic coordinate by a quantity in $[-ms, ms]$, and so it is itself a geometric
semantic mutation. This is the idea at the basis of the Semantic Learning algorithm with Inflate and deflate Mutations (SLIM_GSGP) that is introduced in
this work. As the name suggests, SLIM_GSGP uses a mixture of two types of
mutation. The first one generates an offspring that is larger than its parent, as
traditional GSM does, and it is called inflate mutation. The other produces an
offspring that is smaller than its parent exploiting the idea explained above, and
it is called deflate mutation. SLIM_GSGP, however, can have different variants,
that are presented in the continuation.

3.1   Variants of SLIM_GSGP

In this work, two different ways of defining a function that takes values
in $[-ms, ms]$ and two different ways of perturbing an individual to obtain ball
mutation on the semantic space are studied.

Different Ways of Defining a Function with Values in $[-ms, ms]$. As
already pointed out in the literature [3,17], it is possible to use only one random
tree, instead of two, to define a function with outputs in $[-ms, ms]$, and this
allows us to save a significant amount of computational resources. The approach
presented in [17] solves the problem using mutation steps extracted from a previously decided distribution. The approach used here is inspired by [3], and it

consists in a normalization of the output of a random expression: given a random expression T_R, an expression that takes values in $[-ms, ms]$ is $ms \cdot (1 - \frac{2}{1+|T_R|})$. Even though, as discussed in [3], for some values of T_R this expression can be not zero-centered (in other words, the distribution of its output values can be biased towards specific directions), still the effectiveness of using this type of normalization with GSM was demonstrated in [3].

Different Ways to Obtain Ball Mutation on the Semantic Space. A small perturbation of a numeric value can trivially be obtained either adding a positive or negative quantity that is close to zero, or multiplying by a quantity that is close to (i.e., either slightly larger or slightly smaller than) one. While traditional GSM uses the first approach, to the best of our knowledge the second approach has never been used so far. In this work, the terms that will be multiplied will be either $1 + ms \cdot (T_{R1} - T_{R2})$ (in case we want to use two random trees) or $1 + ms \cdot (1 - \frac{2}{1+|T_R|})$ (in case we want to use just one random tree). When inflate mutation multiplies by one of these terms, the corresponding deflate mutation divides by a term that was previously multiplied, thus allowing for the simplification of that term, and the consequent generation of an offspring that is smaller than its parent.

Combining these two ways of defining a function with outputs in $[-ms, ms]$ (using one or two random trees) and these two ways of obtaining ball mutation (by summing a quantity close to zero or by multiplying by a quantity close to one), we are able to define four SLIM_GSGP variants: SLIM+2, SLIM+1, SLIM*2 and SLIM*1. The inflate and deflate mutations used by these four SLIM_GSGP variants are defined in the continuation.

SLIM+2. The inflate mutation of SLIM+2 is exactly identical to the GSM defined in Sect. 2. Its deflate mutation is defined as follows.

Deflate Mutation of SLIM+2. Given a parent function $T + \sum_{i=1}^{n} ms \cdot (T_{Ri} - T_{Ri+1})$, with $n > 0$, the deflate mutation first draws an index j at random with uniform probability from $\{1, 2, ..., n\}$, and then returns the function $T + \sum_{i=1}^{n} ms \cdot (T_{Ri} - T_{Ri+1}) - ms \cdot (T_{Rj} - T_{Rj+1})$, or, in other terms, the function $T + \sum_{i=1}^{j-1} ms \cdot (T_{Ri} - T_{Ri+1}) + \sum_{i=j+1}^{n} ms \cdot (T_{Ri} - T_{Ri+1})$.

Exactly like for the case of GSM, in this work the two random trees used in the inflate mutation of SLIM+2 will be wrapped around a logistic function, in order to constraint their values in $[0, 1]$.

SLIM+1. The difference between SLIM+1 and SLIM+2 is that SLIM+1 uses one random tree to define a random function that takes values in $[-ms, ms]$. Its inflate and deflate mutations are defined as follows:

Inflate Mutation of SLIM+1. Given a parent function T, this operator returns the function $T + ms \cdot (1 - \frac{2}{1+|T_R|})$, where T_R is a random function and ms is the mutation step.

Deflate Mutation of SLIM+1. Given a parent function $T + \sum_{i=1}^{n} ms \cdot (1 - \frac{2}{1+|T_{Ri}|}))$, with $n > 0$, the deflate mutation first draws an index j at random with uniform probability from $\{1, 2, ..., n\}$, and then returns the function $T + \sum_{i=1}^{j-1} ms \cdot (1 - \frac{2}{1+|T_{Ri}|}) + \sum_{i=j+1}^{n} ms \cdot (1 - \frac{2}{1+|T_{Ri}|})$.

SLIM*2. The difference between SLIM*2 and SLIM+2 is that SLIM*2 uses multiplication by a value close to one to obtain ball mutation in the semantic space. The genetic operators used by SLIM*2 are:

*Inflate Mutation of SLIM*2.* Given a parent function T, this operator returns the function $T \cdot (1 + ms \cdot (T_{R1} - T_{R2}))$, where T_{R1} and T_{R2} are random functions and ms is the mutation step.

*Deflate Mutation of SLIM*2.* Given a parent function $T \cdot \prod_{i=1}^{n}(1 + ms \cdot (T_{Ri} - T_{Ri+1}))$, with $n > 0$, the deflate mutation first draws an index j at random with uniform probability from $\{1, 2, ..., n\}$, and then returns the function $T \cdot \prod_{i=1}^{j-1}(1 + ms \cdot (T_{Ri} - T_{Ri+1})) \cdot \prod_{i=j+1}^{n}(1 + ms \cdot (T_{Ri} - T_{Ri+1}))$.

Also for SLIM*2 the two random trees used in the inflate mutation will be wrapped around a logistic function, in order to constraint their values in $[0, 1]$.

SLIM*1. The difference between SLIM*1 and SLIM*2 is that SLIM*1 uses one random tree to define a random function that takes values in $[-ms, ms]$. Its inflate and deflate mutations are defined as follows:

*Inflate Mutation of SLIM*1.* Given a parent function T, this operator returns the function $T \cdot (1 + ms \cdot (1 - \frac{2}{1+|T_R|}))$, where T_R is a random function and ms is the mutation step.

*Deflate Mutation of SLIM*1.* Given a parent function $T \cdot \prod_{i=1}^{n}(1 + ms \cdot (1 - \frac{2}{1+|T_{Ri}|}))$, with $n > 0$, the deflate mutation first draws an index j at random with uniform probability from $\{1, 2, ..., n\}$, and then returns the function $T \cdot \prod_{i=1}^{j-1}(1 + ms \cdot (1 - \frac{2}{1+|T_{Ri}|})) \cdot \prod_{i=j+1}^{n}(1 + ms \cdot (1 - \frac{2}{1+|T_{Ri}|}))$.

From the previous definitions, it is clear that, for all the SLIM_GSGP variants, the deflate mutation can only be applied to individuals that have already undergone at least one inflate mutation event.

Linked-List Implementation of SLIM_GSGP. In this work, the SLIM_GSGP algorithms have been implemented by representing the genotypes as linked lists. In this way, an inflate mutation can be implemented by a simple append of an item at the end of the list, while a deflate mutation can be implemented by a simple delete of one of the items of the list. Finally, an individual can be evaluated simply accumulating the sum (for SLIM+1 and SLIM+2) or the product (for SLIM*1 and SLIM*2) of the terms in the list. A simple example of a SLIM+2 genotype and the individuals resulting from the application of a sequence of an inflate and a deflate mutation is represented in Fig. 1. As we can observe, the head of the list contains the initial tree T, and it is created at the initialization of the population. Inflate mutation can always be applied, and it simply appends a new list item. As it is clear from the previous definitions, deflate mutation can be applied only if the length of the list is strictly larger than one (in other words, deflate mutation can be applied only to individuals

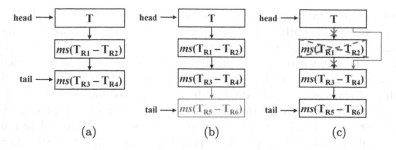

Fig. 1. Part (a) represents an example genotype of a SLIM+2 individual. Part (b) represents an individual resulting from the application of an inflate mutation to the previous one. Part (c) represents an individual resulting from the application of a deflate mutation.

that, besides the head of the list, contain at least another item) and it deletes one random item that is different from the head. For this, it should be clear that the deflate mutation is not a backtracking operator. In fact, deflate mutation does not have to necessarily remove the last element of the list, but can remove any of them, except the head.

4   Test Problems and Experimental Settings

The test problems used in this work are six real-life symbolic regression applications that have often been used as benchmarks in the GP literature (see for instance [2, 9, 10, 26, 32]): Toxicity (626 features and 274 samples, see [2]); Instanbul (7 features and 536 samples, see [1]); Energy (8 features and 330 samples, randomly selected from the dataset presented in [27]); PPB (626 features and 234 samples, see [2]); Concrete (8 features and 1030 samples, see [9]) and Residential Buildings Sales Price (RBSP) (107 features and 372 samples, see [25]). On these problems, we compared standard GP (stdGP) and GSGP to SLIM+2, SLIM+1, SLIM*2 and SLIM*1. A total of 30 runs were performed with each one of these techniques, each run using a different randomly generated partition of the dataset into training (70%) and test (30%) sets. This widely used method is also called *Monte Carlo* cross-validation [11, 33]. All the studied GP variants used a population of 100 individuals, evolved for 2000 generations (as in [31]). Individuals were built using a primitive function set containing the four binary arithmetic operators $+$, $-$, $*$, and $/$ protected as in [15]. Fitness was calculated as the root mean squared error (RMSE) between predicted and expected outputs. The terminal set contained a number of variables corresponding to the number of features in each dataset. The population was initialized using the ramped half-and-half algorithm with a maximum tree depth equal to 6, selection was performed using tournament of size equal to 2 and elitism was used, consisting in the copy of the best individual of the population unchanged into the next generation. stdGP used standard subtree swapping crossover [15] and standard subtree mutation [15], respectively with a rate of 0.8 and 0.2. The maximum tree depth during the evolution was equal to 17 [15]. It is noteworthy that past research has demonstrated that eliminating any restrictions on tree depth during evolution results in significant overfitting of stdGP across various examined test problems [34]. Therefore, in addition to being a conventional and widely adopted approach, the imposition of a depth limitation serves as a regularization technique, providing stdGP with a potential advantage in terms of its generalization capability. Concerning SLIM_GSGP, it is important to emphasize that the application of inflate and deflate mutations does not occur in a sequential manner at each iteration. In other words, it is not a process where the inflate mutation is first applied, followed by the corresponding deflate mutation to generate an offspring. Instead, in all the SLIM_GSGP variants these two mutations are considered distinct and entirely independent operators, each applied with its own probability. More specifically, for the Toxicity dataset all the SLIM_GSGP variants used an inflate mutation rate of 0.1 and a deflate mutation rate of 0.9, with a mutation step randomly generated at each inflate mutation event with uniform probability from $[0, 0.1]$. For the Concrete dataset all the SLIM_GSGP variants used an inflate mutation rate of 0.5 and a deflate mutation rate of 0.5, with a mutation step randomly generated at each inflate mutation event with uniform probability from $[0, 0.3]$. For all the other test problems, all the SLIM_GSGP variants used an inflate mutation rate of 0.3 and a deflate mutation rate of 0.7, with a mutation step randomly generated at each inflate mutation event with uniform probability from $[0, 1]$. Notice that using a different random mutation step at each mutation event was recommended, for instance, in [34]. GSGP used a GSM rate equal to 1, a crossover rate equal to 0 (as suggested in [21, 34]) and a random mutation step, generated at each mutation event with uniform probability from $[0, 1]$, as in [34].

5   Experimental Results

Evolution of the Root Mean Squared Error. The experimental results are reported in Fig. 2, using curves of the fitness (RMSE) on the training and test sets. For each generation, the training fitness of the best individual in the population, as well as its fitness in the test set (that we call test fitness) were recorded. The curves in the plots report the median of these values over the 30 runs. Finally, to verify the statistical significance of the results, Table 1 reports the p-values returned by the Wilcoxon rank-sum test with a significance level $\alpha = 0.05$ and with Bonferroni correction, under the alternative hypothesis that the samples do not have equal medians. In this table, the results indicating a statistically significant difference are highlighted in bold and the cases in which SLIM_GSGP outperforms its competitor in a statistically significant way are labelled with a checkmark ($\checkmark$). We can observe that, while the SLIM_GSGP variants have a generally slower optimization on the training set, for each one of the studied test problems there is at least one SLIM_GSGP variant that either outperforms or performs comparably to stdGP and GSGP on the test set.

Evolution of the Program Size. Figure 3 reports the evolution of the median of the size (i.e., number of tree nodes) of the best individual in the population,

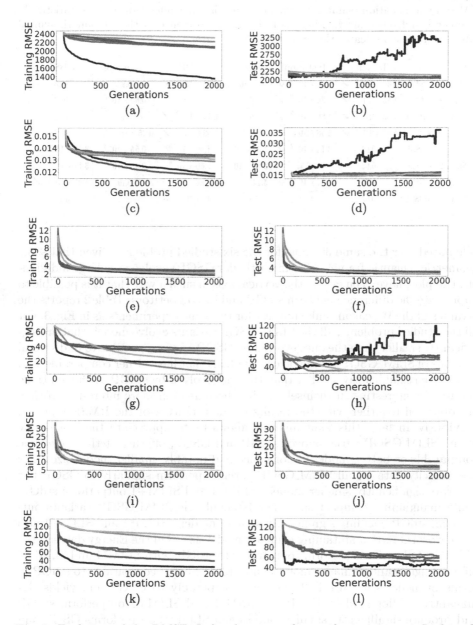

Fig. 2. Evolution of the RMSE. Left column: results on the training set. Right column: results on the test set. Plots (a) and (b): Toxicity. Plots (c) and (d): Instanbul. Plots (e) and (f): Energy. Plots (g) and (h): PPB. Plots (i) and (j): Concrete. Plots (k) and (l): RBSP. For all the plots, the legend is:

STDGP   GSGP   SLIM*1   SLIM*2   SLIM+1   SLIM+2

Table 1. p-values returned by the Wilcoxon rank-sum test for the evolution of the RMSE at termination (same experiments as in Fig. 2). Values indicating a statistically significant difference are highlighted in bold. The presence of the checkmark symbol after the p-value indicates that SLIM_GSGP is better.

	Toxicity	Instanbul	Energy	PPB	Concrete	RBSP
SLIM*1 vs stdGP	**2.60 × 10⁻⁸** ✓	**1.86 × 10⁻⁹** ✓	0.16	**0.13 × 10⁻³** ✓	0.23	0.28
SLIM*2 vs stdGP	**3.85 × 10⁻⁷** ✓	**1.30 × 10⁻⁸** ✓	0.22	0.11	**0.50 × 10⁻³**	0.29
SLIM+1 vs stdGP	**1.63 × 10⁻⁷** ✓	**1.86 × 10⁻⁹** ✓	0.12	**0.14 × 10⁻²** ✓	0.12	0.42
SLIM+2 vs stdGP	**8.00 × 10⁻⁸** ✓	**5.58 × 10⁻⁹** ✓	0.90	**7.99 × 10⁻⁶**	**0.95 × 10⁻³** ✓	**0.55 × 10⁻³**
SLIM*1 vs GSGP	0.62	**3.14 × 10⁻⁷** ✓	**1.39 × 10⁻⁵**	**1.86 × 10⁻⁹**	**0.55 × 10⁻³**	**2.76 × 10⁻⁶** ✓
SLIM*2 vs GSGP	0.62	0.55	**8.85 × 10⁻⁵**	**2.60 × 10⁻⁸**	**2.34 × 10⁻⁶**	**3.14 × 10⁻⁷** ✓
SLIM+1 vs GSGP	0.74	**4.42 × 10⁻⁶** ✓	**5.14 × 10⁻⁶**	**1.86 × 10⁻⁹**	0.10	**1.99 × 10⁻⁶** ✓
SLIM+2 vs GSGP	0.02	0.018	0.38	**0.40 × 10⁻²** ✓	0.85	0.012

calculated over the same 30 runs, for the six studied problems. Given that some geometric semantic frameworks (in particular GSGP) evolve very large individuals, the plots have been cut on the vertical axis in such a way that it is possible to appreciate the difference between stdGP and its competitors. Table 2 reports the p-values of the Wilcoxon rank-sum test for the same experiments as in Fig. 3. For all the studied problems, all the SLIM_GSGP variants evolve models that are significantly smaller than the ones evolved by GSGP and it is possible to find several versions of SLIM_GSGP that evolve individuals that have either comparable size, or are even significantly smaller than the ones evolved by stdGP. Although these results are interesting in themselves, they become even more interesting if they are observed together with the results of the evolution of the RMSE reported previously. In fact, this joint analysis allows us to appreciate the superiority of the SLIM_GSGP variants over stdGP and GSGP, offering at the same time comparable or better model performance and notably smaller model sizes. For the Toxicity dataset, all SLIM_GSGP variants perform similarly to GSGP but generate significantly smaller models. SLIM*1 and SLIM+1 outperform stdGP, while maintaining compact sizes. For Istanbul, the SLIM_GSGP variants outperform stdGP, yielding significantly smaller models. SLIM*1 and SLIM+1 also outperform GSGP, maintaining compact model sizes. On the Energy dataset, the SLIM_GSGP variants perform comparably to stdGP, with SLIM*1 and SLIM+1 offering significantly smaller models and SLIM*2 being comparable to stdGP in terms of model size. SLIM+2 performs comparably to GSGP and yields significantly smaller models. For PPB, SLIM*1 and SLIM+1 outperform stdGP and produce significantly smaller models and SLIM+2 outperforms GSGP and generates significantly smaller models. On the Concrete problem, SLIM*1 and SLIM+1 surpass stdGP, while maintaining significantly smaller model sizes. SLIM+1 and SLIM+2 perform comparably to GSGP, both generating significantly smaller models. Lastly, for RBSP, SLIM*1, SLIM*2 and SLIM+1 perform comparably to stdGP, with SLIM*1 and SLIM+1 producing significantly smaller

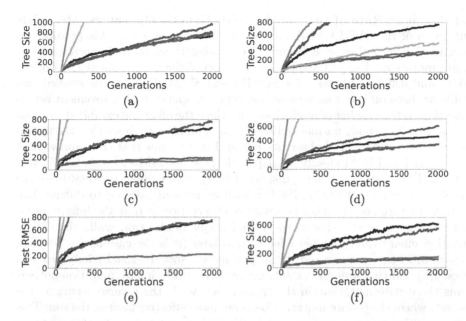

Fig. 3. Evolution of the program size, measured as the median number of nodes of the best individual on the training set. Plots (a): Toxicity. Plots (b): Instanbul. Plots (c): Energy. Plots (d): PPB. Plots (e): Concrete. Plots (f): RBSP. For all the plots, the legend is:

STDGP   GSGP   SLIM*1   SLIM*2   SLIM+1   SLIM+2

Table 2. p-values returned by the Wilcoxon rank sum-test for the evolution of the program size at termination (same experiments as in Fig. 3). Values indicating a statistically significant difference are highlighted in bold. The presence of the checkmark symbol after the p-value indicates that SLIM_GSGP is better (i.e., the individuals evolved by SLIM_GSGP are smaller).

	Toxicity	Instanbul	Energy	PPB	Concrete	RBSP
SLIM*1 vs stdGP	0.091	1.86×10^{-9} ✓	1.86×10^{-9} ✓	0.12×10^{-3} ✓	1.86×10^{-9} ✓	1.86×10^{-9} ✓
SLIM*2 vs stdGP	0.13×10^{-3}	1.86×10^{-9} ✓	0.12	0.16×10^{-3}	0.51	0.049
SLIM+1 vs stdGP	0.29	1.86×10^{-9} ✓	1.86×10^{-9} ✓	0.20×10^{-3} ✓	1.86×10^{-9} ✓	1.86×10^{-9} ✓
SLIM+2 vs stdGP	1.86×10^{-9}	1.02×10^{-7} ✓	1.86×10^{-9}	1.86×10^{-9}	1.86×10^{-9}	1.86×10^{-9}
SLIM*1 vs GSGP	1.86×10^{-9} ✓	1.86×10^{-9} ✓	1.86×10^{-9} ✓	1.86×10^{-9} ✓	1.86×10^{-9} ✓	1.86×10^{-9} ✓
SLIM*2 vs GSGP	1.86×10^{-9} ✓	3.72×10^{-9} ✓	1.86×10^{-9} ✓	1.86×10^{-9} ✓	1.86×10^{-9} ✓	1.86×10^{-9} ✓
SLIM+1 vs GSGP	1.86×10^{-9} ✓	1.86×10^{-9} ✓	1.86×10^{-9} ✓	1.86×10^{-9} ✓	1.86×10^{-9} ✓	1.86×10^{-9} ✓
SLIM+2 vs GSGP	1.86×10^{-9} ✓	1.86×10^{-9} ✓	1.86×10^{-9} ✓	1.86×10^{-9} ✓	1.86×10^{-9} ✓	1.86×10^{-9} ✓

models and SLIM*2 producing models of comparable size to stdGP. Additionally, SLIM*1, SLIM*2, and SLIM+1 outperform GSGP, while SLIM+2 performs comparably to GSGP, all with significantly smaller model sizes.

Improvement Rate of the SLIM_GSGP Genetic Operators. The deflate mutations of the SLIM_GSGP variants are novel operators. This prompts an investigation into their effectiveness, comparing them to their corresponding inflate mutations. Figure 4 shows the evolution of the improvement rates of the inflate and deflate mutations for SLIM*1 and SLIM+1 on all the studied test problems, both on the training and test sets. An operator's improvement rate is quantified as the number of applications in which the offspring exhibited superior fitness than its parent, normalized by the total number of times that the operator was applied. An initial observation from Fig. 4 reveals that the improvement rate curves for SLIM*1 closely overlap with those of SLIM+1 (and also SLIM*2 and SLIM+2, despite not being displayed here). This enables us to evaluate the overall effectiveness of SLIM_GSGP operators without needing to differentiate between its variants. Another noteworthy observation is that the inflate mutation is often more effective in the initial stages of the run, while the deflate mutation often exhibits higher effectiveness later (it is the case of the Toxicity, Instanbul, PPB, Concrete and RBSP problems). Notably, the Toxicity problem (Figs. 4(a) and (b)) presents an intriguing pattern: the inflate mutation outperforms the deflate mutation on the training set, while the reverse occurs on the test set, where the deflate mutation becomes more effective later in the run. This pattern, although most pronounced in Toxicity, is not unique to this problem. Two other test problems (Instanbul and PPB) exhibit a more pronounced difference in the effectiveness of the two operators on the test set, with the deflate mutation proving more effective on the test set than on the training set. This observation suggests that the deflate mutation serves as an implicit regularization mechanism, enabling models to fit less to noisy data points. Consequently, it is unsurprising that the differences between the operators on the training and test sets are less pronounced for datasets that are more consistent and homogeneous i.e., in which similar observations tend to have similar target values (Energy, Concrete, and RBSP). These findings suggest that relying solely on RMSE on the training set as a fitness function may not be advisable, especially for problems vulnerable to overfitting. Instead, a more promising approach for the future could involve multiobjective optimization, with size and/or functional complexity optimized alongside RMSE. This aligns with recommendations often found in the literature [12,14,22,35].

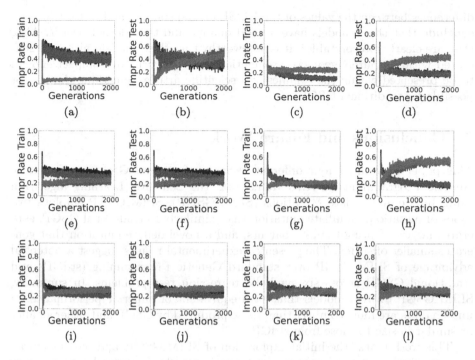

Fig. 4. Evolution of the improvement rate of the mutation operators of SLIM*1 and SLIM+1. First and third columns: results on the training set. Second and fourth columns: results on the test set. Plots (a) and (b): Toxicity. Plots (c) and (d): Instanbul. Plots (e) and (f): Energy. Plots (g) and (h): PPB. Plots (i) and (j): Concrete. Plots (k) and (l): RBSP. For all the plots, the legend is:

━━━ inflate*1   ━━━ deflate*1   ━━━ inflate+1   ━━━ deflate+1

5.1 Example Models Evolved by SLIM_GSGP

Let ms be the mutation step and let: $f(x) = ms \cdot \left(1 - \frac{2}{1+|x|}\right)$. Four models evolved by SLIM+1 on the Instanbul dataset, arbitrarily chosen among the individuals obtained at the end of the 30 runs, are:

$$\text{EM} + \text{FTSE} + f\left(\text{EM} - \left(\text{FTSE} - \left(\tfrac{\text{FTSE}}{\text{FTSE}} - \text{BOVESPA}\right)\right)\right) \tag{3}$$

$$\text{EM} \cdot \text{DAX} + \text{EU} + f\left(\tfrac{\text{EU}}{\text{EU}} + \text{EM}\right) \tag{4}$$

$$\text{EU} + \text{EM} \cdot \text{SP} + \tfrac{\text{NIKKEI} \cdot (\text{BOVESPA}+\text{SP}) \cdot \text{NIKKEI}}{\text{EM}} + f\left(\text{NIKKEI} + \tfrac{\text{SP}}{\text{SP}}\right) + f\left(\text{EM} - \text{FTSE} + \tfrac{\text{NIKKEI}}{\text{NIKKEI}}\right) \tag{5}$$

$$\text{EU} + \text{NIKKEI} + f\left(\text{NIKKEI} - \tfrac{\text{DAX}}{\text{DAX}}\right) + f\left(\tfrac{\text{EU}}{\text{EU}} + \text{EM}\right) \tag{6}$$

The model in Eq. (3) (respectively, Eq. (4, (5 and (6) has an RMSE on the training set of 0.0140 (respectively, 0.0142, 0.0149 and 0.0153), and an RMSE on the test set of 0.0148 (respectively, 0.0147, 0.0146 and 0.0142). Comparing these values with the ones reported in Fig. 2(c) and (d), and observing the small differences between the values of the RMSE on the training and test set, we can conclude that these models have a good quality and they do not overfit. Also, they are clearly interpretable[2]. It is noteworthy that these models, although generated within a purely geometric semantic GP framework, are compact enough to be presented within the confines of a scientific article, something that is not possible for traditional GSGP.

6   Conclusions and Future Work

This paper introduced four different variants of a novel Geometric Semantic Genetic Programming (GSGP) framework, the Semantic Learning algorithm using Inflate and deflate Mutations (SLIM_GSGP). SLIM_GSGP employs two types of mutation: an inflate mutation that, similarly to traditional GSGP, generates larger offspring than its parents, and a novel deflate mutation that generates smaller offspring. The presented experimental results suggest a potential advantage of SLIM_GSGP over standard Genetic Programming (stdGP) and traditional GSGP across six symbolic regression test problems. In summary, SLIM_GSGP evolves models that have equivalent or superior performance on unseen data, are more compact than those produced by GSGP and either smaller or similar in size to those from stdGP.

This work marks the initial exploration of SLIM_GSGP, and paves the way to promising substantial avenues for future research. A comprehensive examination of the hyperparameters governing the diverse SLIM_GSGP variants is a crucial step. Enhancing the fitness function beyond the simple use of the RMSE on the training set holds great potential. Multiobjective optimization, particularly the co-optimization of model size and functional complexity, is a natural candidate. Optimizing the application of the deflate mutation also presents an intriguing prospect. This entails studying strategic time instants for its use and implementing more deliberate subtree selection for removal from parent genotypes, as opposed to the current random pruning approach. Furthermore, we are contemplating the introduction of innovative crossover operators, capable of generating offspring smaller than their parents.

Acknowledgments. This work was supported by national funds through FCT (Fundação para a Ciência e a Tecnologia), under the project - UIDB/04152/2020 - Centro de Investigação em Gestão de Informação (MagIC)/NOVA IMS.

References

1. Akbilgic, O., Bozdogan, H., Balaban, M.E.: A novel hybrid RBF neural networks model as a forecaster. Stat. Comput. **24**(3), 365–375 (2014). https://doi.org/10.1007/s11222-013-9375-7

---

[2] For an explanation of the input variables of the Instanbul dataset (that represent stock market indicators), the reader is referred to [1].

2. Archetti, F., Lanzeni, S., Messina, E., Vanneschi, L.: Genetic programming for computational pharmacokinetics in drug discovery and development. Genet. Program Evolvable Mach. **8**(4), 413–432 (2007)
3. Bakurov, I., et al.: Geometric semantic genetic programming with normalized and standardized random programs. Genet. Program Evolvable Mach. **25**, 6 (2024). https://doi.org/10.1007/s10710-024-09479-1
4. Castelli, M., Manzoni, L.: GSGP-C++ 2.0: a geometric semantic genetic programming framework. SoftwareX **10**, 100313 (2019). https://doi.org/10.1016/j.softx.2019.100313. https://www.sciencedirect.com/science/article/pii/S2352711019301736
5. Castelli, M., Manzoni, L., Gonçalves, I., Vanneschi, L., Trujillo, L., Silva, S.: An analysis of geometric semantic crossover: a computational geometry approach. In: International Joint Conference on Computational Intelligence (2016)
6. Castelli, M., Trujillo, L., Vanneschi, L.: Energy consumption forecasting using semantic-based genetic programming with local search optimizer. Intell. Neurosci. **2015** (2015). https://doi.org/10.1155/2015/971908
7. Castelli, M., Trujillo, L., Vanneschi, L., Silva, S., Z-Flores, E., Legrand, P.: Geometric semantic genetic programming with local search. In: Proceedings of the 2015 Annual Conference on Genetic and Evolutionary Computation, GECCO 2015, pp. 999–1006. Association for Computing Machinery, New York (2015). https://doi.org/10.1145/2739480.2754795
8. Castelli, M., Vanneschi, L., Popovič, A.: Controlling individuals growth in semantic genetic programming through elitist replacement. Intell. Neurosci. **2016** (2016). https://doi.org/10.1155/2016/8326760
9. Castelli, M., Vanneschi, L., Silva, S.: Prediction of high performance concrete strength using genetic programming with geometric semantic genetic operators. Expert Syst. Appl. **40**(17), 6856–6862 (2013)
10. Castelli, M., Vanneschi, L., Silva, S., Ruberto, S.: How to exploit alignment in the error space: two different GP models. In: Riolo, R., Worzel, W.P., Kotanchek, M. (eds.) Genetic Programming Theory and Practice XII. GEC, pp. 133–148. Springer, Cham (2015). https://doi.org/10.1007/978-3-319-16030-6_8
11. Dubitzky, W., Granzow, M., Berrar, D.P.: Fundamentals of Data Mining in Genomics and Proteomics. Springer, Cham (2006)
12. Galván, E., Schoenauer, M.: Promoting semantic diversity in multi-objective genetic programming. In: Proceedings of the Genetic and Evolutionary Computation Conference, GECCO 2019, pp. 1021–1029. Association for Computing Machinery, New York (2019). https://doi.org/10.1145/3321707.3321854
13. Gonçalves, I., Silva, S., Fonseca, C.M.: On the generalization ability of geometric semantic genetic programming. In: Machado, P., et al. (eds.) EuroGP 2015. LNCS, vol. 9025, pp. 41–52. Springer, Cham (2015). https://doi.org/10.1007/978-3-319-16501-1_4
14. Kommenda, M., Kronberger, G., Affenzeller, M., Winkler, S.M., Burlacu, B.: Evolving simple symbolic regression models by multi-objective genetic programming. In: Riolo, R., Worzel, B., Kotanchek, M., Kordon, A. (eds.) Genetic Programming Theory and Practice XIII. GEC, pp. 1–19. Springer, Cham (2016). https://doi.org/10.1007/978-3-319-34223-8_1
15. Koza, J.R.: Genetic Programming: On the Programming of Computers by Means of Natural Selection. MIT Press, Cambridge (1992)
16. Martins, J.F.B.S., Oliveira, L.O.V.B., Miranda, L.F., Casadei, F., Pappa, G.L.: Solving the exponential growth of symbolic regression trees in geometric seman-

tic genetic programming. In: Proceedings of the Genetic and Evolutionary Computation Conference, GECCO 2018, pp. 1151–1158. Association for Computing Machinery, New York (2018). https://doi.org/10.1145/3205455.3205593

17. McDermott, J., Agapitos, A., Brabazon, A., O'Neill, M.: Geometric semantic genetic programming for financial data. In: Esparcia-Alcázar, A.I., Mora, A.M. (eds.) EvoApplications 2014. LNCS, vol. 8602, pp. 215–226. Springer, Heidelberg (2014). https://doi.org/10.1007/978-3-662-45523-4_18

18. Moraglio, A.: An efficient implementation of GSGP using higher-order functions and memoization. In: Johnson, C., Krawiec, K., Moraglio, A., O'Neill, M. (eds.) Semantic Methods in Genetic Programming, Ljubljana, Slovenia (2014). http://www.cs.put.poznan.pl/kkrawiec/smgp2014/uploads/Site/Moraglio2.pdf. Workshop at Parallel Problem Solving from Nature 2014 Conference

19. Moraglio, A., Krawiec, K., Johnson, C.G.: Geometric semantic genetic programming. In: Coello, C.A.C., Cutello, V., Deb, K., Forrest, S., Nicosia, G., Pavone, M. (eds.) PPSN 2012. LNCS, vol. 7491, pp. 21–31. Springer, Heidelberg (2012). https://doi.org/10.1007/978-3-642-32937-1_3

20. Moraglio, A., Mambrini, A.: Runtime analysis of mutation-based geometric semantic genetic programming for basis functions regression. In: Proceedings of the 15th Annual Conference on Genetic and Evolutionary Computation, GECCO 2013, pp. 989–996. Association for Computing Machinery, New York (2013). https://doi.org/10.1145/2463372.2463492

21. Moraglio, A., Mambrini, A.: Runtime analysis of mutation-based geometric semantic genetic programming for basis functions regression. In: Proceedings of the Annual International Conference on Genetic and Evolutionary Computation, GECCO 2013, pp. 989–996. ACM, New York (2013)

22. Parrott, D., Li, X., Ciesielski, V.: Multi-objective techniques in genetic programming for evolving classifiers. In: 2005 IEEE Congress on Evolutionary Computation, vol. 2, pp. 1141–1148 (2005). https://doi.org/10.1109/CEC.2005.1554819

23. Pawlak, T.P., Krawiec, K.: Competent geometric semantic genetic programming for symbolic regression and Boolean function synthesis. Evol. Comput. 26(2), 177–212 (2018)

24. Pietropolli, G., Manzoni, L., Paoletti, A., Castelli, M.: Combining geometric semantic GP with gradient-descent optimization. In: Medvet, E., Pappa, G., Xue, B. (eds.) EuroGP 2022. LNCS, vol. 13223, pp. 19–33. Springer, Cham (2022). https://doi.org/10.1007/978-3-031-02056-8_2

25. Rafiei, M.: Residential Building Data Set. UCI Machine Learning Repository (2018). https://doi.org/10.24432/C5S896

26. Silva, S., Vanneschi, L.: Operator equalisation, bloat and overfitting: a study on human oral bioavailability prediction. In: Rothlauf, F. (ed.) Genetic and Evolutionary Computation Conference, GECCO 2009, Proceedings, Montreal, Québec, Canada, 8–12 July 2009, pp. 1115–1122. ACM (2009)

27. Tsanas, A., Xifara, A.: Accurate quantitative estimation of energy performance of residential buildings using statistical machine learning tools. Energy Build. 49, 560–567 (2012). https://doi.org/10.1016/j.enbuild.2012.03.003. https://www.sciencedirect.com/science/article/pii/S037877881200151X

28. Vanneschi, L.: An introduction to geometric semantic genetic programming. In: Schütze, O., Trujillo, L., Legrand, P., Maldonado, Y. (eds.) NEO 2015. SCI, vol. 663, pp. 3–42. Springer, Cham (2017). https://doi.org/10.1007/978-3-319-44003-3_1

29. Vanneschi, L., et al.: Improving maritime awareness with semantic genetic programming and linear scaling: prediction of vessels position based on AIS data. In: Mora, A.M., Squillero, G. (eds.) EvoApplications 2015. LNCS, vol. 9028, pp. 732–744. Springer, Cham (2015). https://doi.org/10.1007/978-3-319-16549-3_59

30. Vanneschi, L., Castelli, M., Gonçalves, I., Manzoni, L., Silva, S.: Geometric semantic genetic programming for biomedical applications: a state of the art upgrade. In: 2017 IEEE Congress on Evolutionary Computation (CEC), pp. 177–184 (2017). https://doi.org/10.1109/CEC.2017.7969311

31. Vanneschi, L., Castelli, M., Manzoni, L., Silva, S.: A new implementation of geometric semantic GP and its application to problems in pharmacokinetics. In: Krawiec, K., Moraglio, A., Hu, T., Etaner-Uyar, A.Ş, Hu, B. (eds.) EuroGP 2013. LNCS, vol. 7831, pp. 205–216. Springer, Heidelberg (2013). https://doi.org/10.1007/978-3-642-37207-0_18

32. Vanneschi, L., Castelli, M., Scott, K., Trujillo, L.: Alignment-based genetic programming for real life applications. Swarm Evol. Comput. **44**, 840–851 (2019). https://doi.org/10.1016/j.swevo.2018.09.006. https://www.sciencedirect.com/science/article/pii/S2210650218300208

33. Vanneschi, L., Silva, S.: Lectures on Intelligent Systems. Natural Computing Series, Springer, Cham (2023). https://doi.org/10.1007/978-3-031-17922-8

34. Vanneschi, L., Silva, S., Castelli, M., Manzoni, L.: Geometric semantic genetic programming for real life applications. In: Riolo, R., Moore, J.H., Kotanchek, M. (eds.) Genetic Programming Theory and Practice XI. GEC, pp. 191–209. Springer, New York (2014). https://doi.org/10.1007/978-1-4939-0375-7_11

35. Vladislavleva, E.J., Smits, G.F., den Hertog, D.: Order of nonlinearity as a complexity measure for models generated by symbolic regression via pareto genetic programming. IEEE Trans. Evol. Comput. **13**(2), 333–349 (2009). https://doi.org/10.1109/TEVC.2008.926486

Improving Generalization of Evolutionary Feature Construction with Minimal Complexity Knee Points in Regression

Hengzhe Zhang[1]($\boxtimes$), Qi Chen[1], Bing Xue[1], Wolfgang Banzhaf[2], and Mengjie Zhang[1]

[1] Centre for Data Science and Artificial Intelligence and School of Engineering and Computer Science, Victoria University of Wellington, PO Box 600, Wellington 6140, New Zealand
hengzhe.zhang@ecs.vuw.ac.nz

[2] Department of Computer Science and Engineering in the College of Engineering and BEACON Center, Michigan State University, East Lansing, MI 48824, USA

Abstract. Genetic programming-based evolutionary feature construction is a widely used technique for automatically enhancing the performance of a regression algorithm. While it has achieved great success, a challenging problem in feature construction is the issue of overfitting, which has led to the development of many multi-objective methods to control overfitting. However, for multi-objective methods, a key issue is how to select the final model from the front with different trade-offs. To address this challenge, in this paper, we propose a novel minimal complexity knee point selection strategy in evolutionary multi-objective feature construction for regression to select the final model for making predictions. Experimental results on 58 datasets demonstrate the effectiveness and competitiveness of this strategy when compared to eight existing methods. Furthermore, an ensemble of the proposed strategy and existing model selection strategies achieves the best performance and outperforms four popular machine learning algorithms.

Keywords: Knee Point · Multi-criteria Decision-Making · Genetic Programming · Evolutionary Feature Construction · Symbolic Regression

1 Introduction

Evolutionary feature construction is an emerging topic that has achieved significant success in enhancing machine learning pipelines [32]. Formally, evolutionary feature construction methods aim to create a set of features $\Phi(X)$ to improve the learning performance of a machine learning algorithm $\mathcal{A}$ on a dataset (X, Y). Among all evolutionary feature construction methods, genetic programming (GP)-based feature construction is one of the most popular choices because its variable-length, flexible representation is a natural approach for feature construction. However, despite its considerable success, a significant challenge in this area is the problem of overfitting [2].

Since evolutionary algorithms are gradient-free optimization techniques, many works are able to optimize non-differentiable complexity measures to strike a balance between training accuracy and model complexity. Many works use a multi-objective optimization framework to balance the trade-off between learning performance and model size, VC-dimension [10], input-output distance correlation [33], or Rademacher complexity [8]. This paper focuses on multi-objective feature construction using the size of GP trees as the complexity measure because of its simplicity. In simple terms, multi-objective feature construction considers both training accuracy/cross-validation score and tree size in the evaluation and selection process. Finally, a Pareto front with different levels of trade-off is obtained for users to select the appropriate model.

However, when confronted with a front consisting of solutions with varying trade-offs, one key issue is how to select the final model from the set of non-dominated solutions, a problem known as multi-objective decision-making. Most existing approaches use the model with the highest training accuracy [8], but this model may still exhibit significant complexity as it is an extreme point in the front.

In the domain of multi-objective decision-making [7], when no explicit preference is given, a common strategy is to identify a knee point [44]. The knee point is a point where a marginal improvement in one objective results in a substantial degradation in other objectives. Based on this definition, it is evident that multiple knee points may exist within the front.

Existing work in multi-objective GP often selects a single knee point based on the most significant trade-off among solutions in the front [38], which is intuitive when domain knowledge is lacking. However, for evolutionary feature construction, we hypothesize that among all knee points, the one with minimal complexity may provide better generalization performance, aligning with the philosophy of Occam's razor [36]. This hypothesis is based on the idea that a substantial increase in complexity required to improve training accuracy may be indicative of overfitting. Therefore, choosing the knee point with minimal complexity is a sensible option to avoid overfitting.

To clarify this, we present a real-world example of a front in Fig. 1 based on the dataset "OpenML_228", which is a case of severe overfitting when applying some non-dominated solutions to the test set. Here, several knee points exist on the Pareto front. However, the model at the knee point with the largest bend angle, denoted as 'B,' performs poorly, with a test relative squared error (RSE) exceeding 0.48. In contrast, the knee point with minimal complexity, denoted as 'D', demonstrates reasonable performance with a test RSE of approximately 0.16. Moreover, this example shows that for knee points on the left side of knee point 'D', there is a gradual increase in test RSE, as indicated by the color of those points, suggesting that overfitting may occur before reaching the complexity level of the traditional knee point. Therefore, selecting the knee point with the largest bend angle, as in the traditional approach, may not be ideal.

1.1 Goals

In this paper, we propose a minimal complexity knee point selection (MCKP) strategy for selecting the final model in multi-objective GP-based feature con-

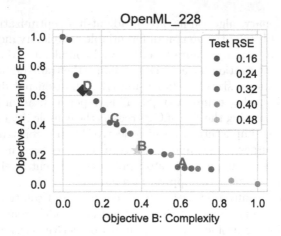

Fig. 1. Visualization of knee points on the front. The numbers in the legend represent relative squared error (RSE) on the test set, with darker colors indicating better performance, and yellow points represent extremely worse RSE. Both objectives are normalized by extreme objective values in the front. Knee points are annotated by red letters. The figure is for post-hoc analysis only and cannot be used for model selection. (Color figure online)

struction for regression[1]. Firstly, our approach uses a clustering algorithm to automatically determine the angle threshold, thereby identifying a set of angle-based knee points. Subsequently, among all knee points, we select the one with minimal complexity as the final model. The main objectives are summarized as follows:

- To favor models with potentially strong generalization performance, we propose a minimal complexity selection strategy to select the final model from the Pareto front.
- To determine a set of candidate knee points from the front, we propose a clustering-based method to automatically determine the angle threshold for knee points.
- To validate the effectiveness of the proposed strategy, we compare the MCKP strategy with seven commonly used model selection strategies in the multi-objective framework on 58 datasets.

1.2   Organization

The remainder of this paper is structured as follows: Sect. 2 reviews related work on knee point selection and overfitting control. Section 3 introduces the details of the proposed algorithm. Section 4 provides the experimental settings, and Sect. 5 shows the experimental results. Section 6 includes further analysis of the proposed strategy. Finally, we conclude the paper and outline future directions in Sect. 7.

---

[1] Source code: https://anonymous.4open.science/r/Knee-GP/.

2   Related Work

2.1   Knee Point Selection

In real-world applications, users often need to select a single solution from the Pareto front as the final solution, which is known as multi-objective decision-making [6]. When no specific preference exists, a common approach is to select the knee point, which is a point where improving one objective significantly decreases another. However, there is no formal, clear definition of a knee point since it depends on the specific context. In machine learning, it could mean the decrease in complexity if the training RSE score improves by 0.1 or 0.01. Due to this ambiguity, several knee point selection strategies exist in the field of multi-objective optimization [30], broadly classified as trade-off information-based and geometry property-based methods [19].

For trade-off information-based knee point selection strategies, a representative example is the utility function. Assuming there are M normalized minimization objectives $f_1, \ldots, f_M$, the trade-off between two points x_i and x_j on the Pareto front can be computed as follows [29]:

$$T\left(x_i, x_j\right) = \frac{\sum_{m=1}^{M} \max\left[0, f_m\left(x_j\right) - f_m\left(x_i\right)\right]}{\sum_{m=1}^{M} \max\left[0, f_m\left(x_i\right) - f_m\left(x_j\right)\right]}$$

This equation calculates the ratio of gain and loss when changing objective values. Then, the utility value of point x_i is defined as the minimum trade-off value $T\left(x_i, x_j\right)$ among all possible x_j on the Pareto front:

$$\mu\left(x_i, S\right) = \min_{x_j \in S} T\left(x_i, x_j\right)$$

Thus, an individual with a high utility value for any changes is considered a knee point [29].

Regarding geometry property-based knee point selection strategies, two representative examples are as follows:

- Angle-based Method [11]: For a bi-objective optimization task, the angle method calculates the angle between the line formed by the current point x and the left point x^L and the line formed by the current point x and the right point x^R. For simplicity, we can first calculate two angles $\theta^L = \arctan \frac{f_2\left(\mathbf{x}^L\right) - f_2(\mathbf{x})}{f_1(\mathbf{x}) - f_1\left(\mathbf{x}^L\right)}$ and $\theta^R = \arctan \frac{f_2(\mathbf{x}) - f_2\left(\mathbf{x}^R\right)}{f_1\left(\mathbf{x}^R\right) - f_1(\mathbf{x})}$, as shown in Fig. 2. The bend angle is then defined as the difference between θ^L and θ^R, i.e., $\theta\left(\mathbf{x}, \mathbf{x}^L, \mathbf{x}^R\right) = \theta^L - \theta^R$. The point with the largest bend angle $\theta\left(\mathbf{x}, \mathbf{x}^L, \mathbf{x}^R\right)$ is chosen as the knee point.
- Distance To Extreme Line [31]: The distance to extreme line method identifies the knee point on the Pareto front by finding the point with the maximum distance from a line $\mathcal{L}\left(p_1^*, p_2^*\right)$, where $\mathcal{L}\left(p_1^*, p_2^*\right)$ represents the line connecting two extreme points p_1^* and p_2^* on the Pareto front.

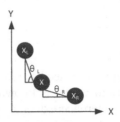

Fig. 2. Angle-based knee point calculation.

In the GP domain, knee point-based selection methods have been used for determining important features [37] and important individuals for knowledge transfer [38]. However, their application in selecting the final model based on the trade-off between training accuracy and model complexity remains limited. This paper explores this aspect.

2.2 Evolutionary Feature Construction

Evolutionary feature construction has been widely used to enhance learning performance and can be categorized into three categories: wrapper-based, filter-based, and embedded methods.

– Wrapper-based methods evaluate features based on a specific learning algorithm, such as KNN [28] and decision trees [43]. These methods can achieve good performance with that specific learning algorithm. However, the wrapper-based method can sometimes lead to overfitting because it directly optimizes accuracy or cross-validation scores on the training data.
– Filter-based methods use general metrics that are independent of any learning algorithm to evaluate features, such as purity [22], which is inexpensive and can generalize to different kinds of algorithms. However, these features may not have optimal performance on a specific learning algorithm.
– Embedded methods construct features during the learning process, with symbolic regression [9] being a typical example.

This paper focuses on wrapper-based methods due to their effectiveness. The problem of overfitting in wrapper-based methods is the issue we aim to address in this paper.

2.3 Overfitting Control for Genetic Programming

GP-based symbolic regression and feature construction methods have achieved great success in recent years. However, a significant challenge in applying evolutionary feature construction in real-world scenarios is its susceptibility to overfitting on limited or noisy training data [1,34]. To address this challenge, various approaches have been explored. Some incorporate metrics from statistical machine learning theory, such as Tikhonov regularization [24], VC dimension [10], or Rademacher complexity [8], as additional optimization objectives. Others adopt overfitting control techniques from other machine learning domains, including auxiliary fitness functions [4], modular architecture [39], semantic hoist mutation [40], multi-task learning [5], feature selection [9], ensemble learning [41], and random sampling [16].

Among these overfitting control techniques, multi-objective GP is widely used in state-of-the-art symbolic regression and evolutionary feature construction algorithms [8,17]. Typically, one objective is set as the training accuracy, and the other objective is the model size. However, a challenge arises in that a front of models with different levels of training performance and complexity is available at the end of evolution. When domain experts are available, they can inspect these models and select the best one. However, in many cases where domain experts are not available, many existing algorithms simply choose the model with the best training accuracy [8,17], which is evidently suboptimal and worth further investigation.

3 The Proposed Method

3.1 Algorithm Framework

Overall, this paper focuses on evolutionary feature construction based on multi-tree GP with a linear regression model. The optimization objectives are leave-one-out cross-validation loss and tree sizes. The evolutionary process follows a common framework of evolutionary feature construction, which includes the following stages:

- Population Initialization: Initially, a set of individuals is randomly initialized using the ramped half-and-half method [3]. Each multi-tree GP individual starts with a single randomly initialized GP tree, representing a constructed feature. Although only one GP tree is initialized in each individual, additional GP trees can be added using genetic operators during offspring generation [20].
- Individual Evaluation: For each individual, the evaluation process first transforms the training data using the features constructed by all trees within a GP individual. Then, a linear regression model is trained on the constructed features to calculate training errors using a leave-one-out cross-validation scheme on the training data [42]. Along with the training error, we also compute the tree size, which is the sum of the sizes of all GP trees within an individual.
- Parent Selection: After obtaining objective values, parents are selected using the domination-based binary tournament selection operator in NSGA-II [12]. The general idea is that, for a pair of randomly selected individuals, the non-dominated solution is given the first priority, and then the individual with the better crowding distance is considered if the two individuals are non-dominated with respect to each other.
- Offspring Generation: Offspring are generated by using random subtree crossover and random subtree mutation on GP trees. Moreover, the random

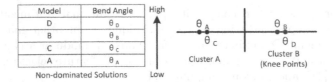

Fig. 3. Automatic determination of the knee point threshold through clustering.

tree addition and random tree deletion operators [20] are used to enable the construction of more than one feature. The crossover operator, mutation operator, and addition/deletion operator are applied sequentially with their respective probabilities.

- Environmental Selection: In this stage, non-dominated sorting with crowding distance [12] is used to select surviving individuals from a combination of parent and offspring individuals.

The evaluation, parent selection, offspring generation, and environmental selection are performed iteratively until the termination criterion is met, resulting in a front of solutions with various trade-offs between model complexity and training accuracy. Subsequently, we can use the knee point selection strategy to identify the final model from this front. The predictions on unseen data are bounded within the range of the training data to avoid overly large extrapolations because we have seen that bounding the predictions, i.e., using decision trees [35] and k-nearest neighbor [18] as the base learner, can provide good generalization performance.

3.2   Minimal-Complexity Knee Selection

In this paper, we first calculate the bend angle θ_Δ for each point x using its adjacent points x_{i-1} and x_{i+1} in the objective space of the front. Then, we can identify knee points based on a threshold of bend angles. However, using a static threshold is challenging, as finding a fixed value suitable for all datasets is difficult. Thus, as illustrated in Fig. 3, K-Means is applied to automatically determine the threshold by clustering non-dominated solutions into k groups based on angles, where k is a hyperparameter. The cluster corresponding to the largest angle is chosen and denoted by $\mathcal{C}_{max}$, representing the cluster of knee points. Within that cluster $\mathcal{C}_{max}$, the model with the minimum complexity is selected as the final model. If the number of points is less than the required number of clusters, then the point with the largest bend angle is selected. The pseudo-code is presented in Algorithm 1, which primarily consists of two stages:

- Angle Calculation (Lines 2–7): The angles are calculated based on the bend angle calculation method introduced in Sect. 2.1.
- Minimal-Complexity Knee Selection (Lines 9–15): After using clustering techniques to determine the threshold for knee points, the knee point with minimal complexity is chosen as the final model.

Algorithm 1. Minimal Complexity Knee Selection

Input: x: Points from Pareto front, k: Number of clusters
Output: Φ_{min}: Selected model with minimum complexity

1: $\Theta \leftarrow \{\}$ ▷ Initialize angle set
2: **for** $i = 1$ to $len(\mathbf{x}) - 1$ **do**   ▷ Iterate through points
3:      $\theta_L \leftarrow \arctan\left(\frac{\mathbf{x}[i-1][1]-\mathbf{x}[i][1]}{\mathbf{x}[i][0]-\mathbf{x}[i-1][0]}\right)$ ▷ Calculate left angle
4:      $\theta_R \leftarrow \arctan\left(\frac{\mathbf{x}[i][1]-\mathbf{x}[i+1][1]}{\mathbf{x}[i+1][0]-\mathbf{x}[i][0]}\right)$ ▷ Calculate right angle
5:      $\theta_\Delta \leftarrow \theta_L - \theta_R$ ▷ Compute the bend angle
6:      $\Theta \leftarrow \Theta \cup \{\theta_\Delta\}$
7: **end for**
8: $\mathcal{C} \leftarrow \text{KMeans}(\Theta, k)$ ▷ Cluster the angles
9: $\theta_{max} \leftarrow -\infty$
10: **for** $\mathcal{C}_j$ in $\mathcal{C}$ **do**   ▷ Find cluster with max angle
11:      **if** $\frac{1}{|\mathcal{C}_j|}\sum_{\theta \in \mathcal{C}_j} \theta > \theta_{max}$ **then**
12:          $\theta_{max} \leftarrow \frac{1}{|\mathcal{C}_j|}\sum_{\theta \in \mathcal{C}_j} \theta$
13:          $\mathcal{C}_{max} \leftarrow \mathcal{C}_j$
14:      **end if**
15: **end for**
16: $\Phi_{min} \leftarrow \text{argmin}_{\Phi \in \mathcal{C}_{max}} \text{Complexity}(\Phi)$ ▷ Select features with minimum complexity
17: **return** Φ_{min}

---

4 Experimental Settings

4.1 Datasets

The datasets consist of real-world datasets from the Penn Machine Learning Benchmark (PMLB) [26], which is a curated list of datasets from OpenML. Synthetic datasets are excluded because they are less prone to overfitting. After excluding the synthetic datasets, 58 datasets remains.

4.2 Evaluation Protocol

To obtain reliable results, we conduct 30 independent runs with different random seeds. In each run, to simulate situations where training samples are scarce, only 100 training instances are used as the training data for feature construction [25], and the remaining data are used for testing. To further increase the difficulty, 20 random variables generated from $\mathcal{N}(0,1)$ are appended to the datasets. To eliminate magnitude differences between different datasets, RSE is used to evaluate the performance of a model on test data. After conducting 30 independent runs, the signed rank test at a significance level of 0.05 was used to examine statistical differences among algorithms.

Table 1. Parameter settings for MCKP-GP.

Parameter	Value
Maximal Population Size	100
Number of Generations	50
Crossover and Mutation Rates	0.9 and 0.1
Tree Addition Rate	0.5
Tree Deletion Rate	0.5
Initial Tree Depth	0–3
Maximum Tree Depth	10
Initial Number of Trees	1
Maximum Number of Trees	20
Elitism (Number of Individuals)	1
Functions	+, −, *, AQ, Sqrt, Max, Min, Negative, Abs, ReLU, Gaussian

4.3 Parameter Settings

The parameter settings are shown in Table 1, which are common settings for GP. For instance, the crossover rate is set significantly higher than the mutation rate to facilitate the exchange of building blocks. To prevent zero-division errors, we employ the analytical quotient (AQ) [23], defined as $AQ = \frac{a}{\sqrt{1+(b^2)}}$ for given inputs a and b. We use *ReLU* and *Gaussian* because they have shown good performance in neuroevolution [15]. The range for ephemeral random constants is set to $[-5\tau, 5\tau]$, where τ represents the maximum absolute value of input variables [35].

4.4 Baseline Algorithms

The baseline algorithms include five popular knee point selection strategies:

- Angle Knee Selection (AKS) [6]: AKS identifies the final model by selecting the point with the maximum angle formed by it, its left neighbor, and its right neighbor. The angle calculation method is introduced in Sect. 2.
- Four Angle Knee Selection (FAKS) [6]: FAKS is similar to AKS, but considers the maximum angle formed by four adjacent points instead of two adjacent points to determine the knee.
- Bended Angle Knee Selection (BAKS) [11]: BAKS is similar to AKS, but uses the two extreme points as reference points for angle calculation, rather than using two adjacent points.
- Utility Function Knee Selection (UFKS) [29]: UFKS selects the individual with the highest utility value as the final model. The utility function is introduced in Sect. 2.
- Distance To Extreme Line Knee Selection (DELKS) [44]: In this method, the model with the maximum Euclidean distance from the extreme line is chosen as the final model.

In addition to knee point selection methods, we also compare:

- Best Training Accuracy (BTA) [8,17]: BTA selects the model with the best training accuracy/lowest training error from the front as the final model.
- Best Harmonic Mean Rank (HMR) [14]: HMR ranks models based on the harmonic mean of accuracy rank (r_a) and model size rank (r_m), using the formula $\frac{1}{r_a^{-1}+r_m^{-1}}$. The model with the best harmonic mean rank is chosen as the final model. This method is used for ranking models discovered by different algorithms in the GECCO 2022 symbolic regression competition [14], but it is also applicable for ranking models within a front.
- Standard GP (STD-GP): STD-GP is a standard GP algorithm that does not consider model size as an additional objective.

Except for STD-GP, all model selection methods follow the same multi-objective evolutionary process, differing only in how they select the final model from the Pareto front.

5 Experimental Results

In this section, we validate the effectiveness of the proposed minimal complexity knee point selection strategy by comparing it with other model selection strategies. Additionally, we inspect the Pareto front and conduct parameter sensitivity analysis to further demonstrate the effectiveness of the proposed method.

5.1 Comparison of Model Selection Strategies

In this section, we present experimental results of test RSE when employing different model selection strategies, as detailed in Table 2. There are two points to highlight from the results.

First, the proposed MCKP strategy significantly improves the generalization performance of standard GP on 32 datasets and degrades it on 11 datasets, indicating that MCKP effectively enhances generalization performance. Traditional knee point selection strategies also outperform standard GP to varying degrees. In comparison, BTA improves performance on only one dataset while worsening it on four datasets. Thus, existing methods for selecting the best training performance are not effective for controlling overfitting, and knee point selection strategies are better options.

Second, the experimental results show that using the knee point with minimal complexity outperforms the AKS strategy on 20 datasets and underperforms on 7 datasets. This is an interesting finding because except for MCKP, other knee point selection strategies show similar behaviors to each other, as most of them exhibit similar performance on more than 50 out of 58 datasets. Ideally, it would be great to know which strategy performs well on which dataset, as instance space analysis techniques show [21]. However, it is a very difficult task because overfitting is not only related to the number of instances but also the noise in data, which is an unknown property. Thus, an alternative way is to combine

Table 2. Statistical comparison of test RSE across various model selection strategies. ("+", "~", and "−" indicate that using the method in a row performs better than, similar to, or worse than using the method in a column.)

	AKS	FAKS	BAKS	MEDKS
MCKP	20(+)/31(~)/7(−)	15(+)/36(~)/7(−)	10(+)/37(~)/11(−)	10(+)/40(~)/8(−)
AKS	—	0(+)/58(~)/0(−)	0(+)/56(~)/2(−)	0(+)/54(~)/4(−)
FAKS	—	—	0(+)/55(~)/3(−)	0(+)/56(~)/2(−)
BAKS	—	—	—	0(+)/58(~)/0(−)
MEDKS	—	—	—	—
UFKS	—	—	—	—
HMR	—	—	—	—
BTA	—	—	—	—
	UFKS	HMR	BTA	STD-GP
MCKP	11(+)/37(~)/10(−)	16(+)/30(~)/12(−)	31(+)/15(~)/12(−)	32(+)/15(~)/11(−)
AKS	0(+)/55(~)/3(−)	2(+)/51(~)/5(−)	23(+)/28(~)/7(−)	21(+)/29(~)/8(−)
FAKS	0(+)/56(~)/2(−)	2(+)/52(~)/4(−)	22(+)/28(~)/8(−)	24(+)/26(~)/8(−)
BAKS	0(+)/58(~)/0(−)	3(+)/52(~)/3(−)	24(+)/29(~)/5(−)	24(+)/27(~)/7(−)
MEDKS	0(+)/58(~)/0(−)	3(+)/52(~)/3(−)	24(+)/28(~)/6(−)	26(+)/26(~)/6(−)
UFKS	—	3(+)/52(~)/3(−)	23(+)/29(~)/6(−)	25(+)/27(~)/6(−)
HMR	—	—	24(+)/31(~)/3(−)	27(+)/26(~)/5(−)
BTA	—	—	—	1(+)/53(~)/4(−)

models selected by two strategies and make an ensemble prediction. By doing so, we hope the model can benefit from two models, which will be shown in Sect. 6.

To further analyze the behavior of different selection strategies, we plot both the evolutionary training curve and the corresponding test curve of these selection methods on four representative datasets. The training curves are shown in Fig. 4a, and they reveal that MCKP has a significantly lower training curve compared to other methods. This aligns with our assumption because MCKP favors the simplest knee point, which has higher training error than traditional knee points. However, as shown in Fig. 4b, other strategies may overfit on datasets like "OpenML_228", whereas MCKP handles overfitting well on these datasets. Thus, in practical scenarios where domain knowledge suggests potentially severe overfitting, considering MCKP for model selection can mitigate the risk of overfitting.

5.2   Visualization of Pareto Fronts

In Sect. 1, we introduced the minimal knee point selection method using an example of the final front. Here, we provide more results on various datasets in Fig. 5. The training error and complexity in these figures are normalized according to the best and worst objective values achieved by non-dominated individuals. These results highlight that knee points with the largest bend angles are not good in many cases. For example, on the "OpenML_210" dataset, the traditional knee point, labeled as point "B", has a relatively high RSE of around

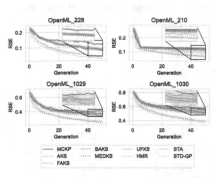

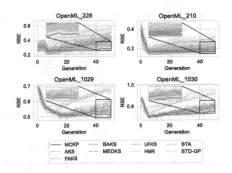

(a) Training RSE of the models selected by different model selection methods.

(b) The corresponding test RSE of the selected models using different model selection methods.

Fig. 4. Evolutionary plots of training RSE and test RSE for the selected models.

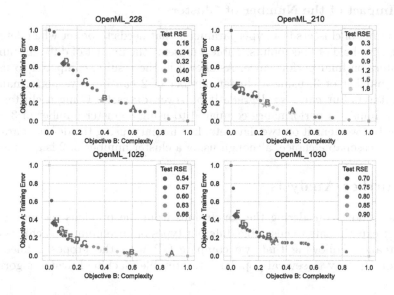

Fig. 5. Visualization of knee points on Pareto fronts. The numbers in the legend represent normalized test MSE, where lower values are better. Knee points are annotated by red letters. The "star" point denotes the traditional knee point, and the "diamond" point represents the minimal complexity knee point. Yellow points represent models with extremely high test errors. (Color figure online)

1.5, whereas the knee point with minimal complexity can achieve a lower RSE of approximately 0.3. Similar trends can also be observed in other figures, suggesting that selecting the knee point with minimal complexity is a better option in many cases.

Table 3. Statistical comparison of test RSE for different numbers of clusters.

	3	5
2	9(+)/42(∼)/7(−)	15(+)/34(∼)/9(−)
3	—	2(+)/55(∼)/1(−)

Table 4. Statistical comparison of test RSE for different model selection strategies

	AKS+HMR	MCKP	AKS	HMR
MCKP+HMR	9(+)/46(∼)/3(−)	12(+)/44(∼)/2(−)	22(+)/35(∼)/1(−)	20(+)/34(∼)/4(−)
AKS+HMR	—	14(+)/38(∼)/6(−)	10(+)/47(∼)/1(−)	4(+)/53(∼)/1(−)
MCKP	—	—	20(+)/31(∼)/7(−)	16(+)/30(∼)/12(−)
AKS	—	—	—	2(+)/51(∼)/5(−)

5.3   Impact of the Number of Clusters

The number of clusters is a hyperparameter that needs to be set when determining the threshold of knee points. Table 3 presents the impact of different numbers of clusters on final performance. As shown in the results, compared to using a cluster number of 3, using a cluster number of 2 can improve performance on nine datasets but can also degrade it on seven datasets. Using a cluster number of 5 improves performance compared to using a cluster number of 3 on one dataset but worsens it on two datasets. In summary, using the default parameter of 3 is a reasonable choice, although using a cluster number of 2 is good as well.

6   Further Analysis

In this section, considering the conclusions from the previous section, we first conduct experiments to ensemble different types of knee points to achieve better performance. Following that, we compare GP with the proposed knee point selection strategy to several popular interpretable machine learning algorithms.

6.1   Post-Hoc Analysis of Ensemble Learning

Even though the MCKP strategy is better than other methods, like HMR, on 16 datasets, it is worth noting that MCKP is still worse than HMR on around 12 datasets. Given that MCKP and other model selection techniques have varying advantages on different datasets, we propose using ensemble learning to enhance performance. In this section, we focus on combining MCKP and HMR because AKS, FAKS, BAKS, MEDKS, and UFKS have similar test RSE to HMR on most datasets, indicating that they select similar models from the front. The experimental results of RSE, presented in Table 4, demonstrate that combining MCKP and HMR through ensemble learning performs better than using MCKP alone on 12 datasets and is worse on only 2 datasets. This indicates that ensemble learning can combine the advantages of two different model selection strategies and

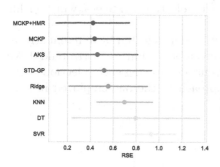

Fig. 6. Median RSE of different learning methods.

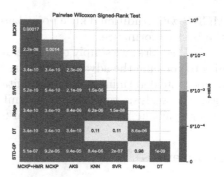

Fig. 7. Pairwise statistical comparison of different learning methods.

achieve better performance. Moreover, we also tried to combine AKS and HMR, but its performance is significantly worse than combining MCKP and HMR on 9 datasets, and only better on 3 datasets. These results demonstrate that it is important to incorporate selection strategies with different behaviors to achieve good performance. AKS and HMR have similar test RSE on 51 datasets, and their similar behavior in model selection results in fewer improvements compared to combining MCKP and HMR.

6.2 Comparisons with Other Machine Learning Algorithms

To further validate the effectiveness of the proposed method, we compare it with popular machine learning algorithms, including support vector regression (SVR), k-nearest neighbor (KNN), Ridge, and decision tree (DT) [27]. The experimental results of RSE are presented in Fig. 6, and the pairwise signed rank test with Benjamini-Hochberg correction is shown in Fig. 7. These results indicate that GP outperforms popular machine learning algorithms when dealing with sample-limited and noisy datasets. Furthermore, the proposed knee point selection strategy further enhances the advantages of GP, especially the ensemble knee point selection strategy.

7   Conclusions

In this paper, we propose a minimal complexity strategy, MCKP, to select the final model from knee points on the front of training accuracy and model size in order to improve the generalization performance of GP-based evolutionary feature construction algorithms. Experimental results on 58 datasets show that MCKP outperforms existing knee point-based model selection strategies and the strategy that selects the model with the best training accuracy/lowest training error in controlling severe overfitting. Given that we usually do not know which dataset is prone to overfitting, we also propose an ensemble strategy, combining

MCKP with traditional knee point-based model selection strategies, which yields the best performance.

In this work, we simply ensemble different models selected by different strategies to combine the advantages of these various models. Future work could delve deeper into determining the most suitable scenario for using each selection strategy. The analysis could follow a similar approach to the one employed in analyzing symbolic regression benchmarks [13].

References

1. Agapitos, A., Loughran, R., Nicolau, M., Lucas, S., O'Neill, M., Brabazon, A.: A survey of statistical machine learning elements in genetic programming. IEEE Trans. Evol. Comput. **23**(6), 1029–1048 (2019)
2. Al-Sahaf, H., et al.: A survey on evolutionary machine learning. J. R. Soc. N. Z. **49**(2), 205–228 (2019)
3. Banzhaf, W., Nordin, P., Keller, R.E., Francone, F.D.: Genetic Programming: An Introduction: On the Automatic Evolution of Computer Programs and Its Applications. Morgan Kaufmann Publishers Inc. (1998)
4. Bi, Y., Xue, B., Zhang, M.: Dual-tree genetic programming for few-shot image classification. IEEE Trans. Evol. Comput. **26**(3), 555–569 (2021)
5. Bi, Y., Xue, B., Zhang, M.: Learning and sharing: a multitask genetic programming approach to image feature learning. IEEE Trans. Evol. Comput. **26**(2), 218–232 (2021)
6. Branke, J., Deb, K., Dierolf, H., Osswald, M.: Finding knees in multi-objective optimization. In: Yao, X., et al. (eds.) PPSN 2004. LNCS, vol. 3242, pp. 722–731. Springer, Heidelberg (2004). https://doi.org/10.1007/978-3-540-30217-9_73
7. Chaudhuri, S., Deb, K.: An interactive evolutionary multi-objective optimization and decision making procedure. Appl. Soft Comput. **10**(2), 496–511 (2010)
8. Chen, Q., Xue, B., Zhang, M.: Rademacher complexity for enhancing the generalization of genetic programming for symbolic regression. IEEE Trans. Cybern. **52**(4), 2382–2395 (2022)
9. Chen, Q., Zhang, M., Xue, B.: Feature selection to improve generalization of genetic programming for high-dimensional symbolic regression. IEEE Trans. Evol. Comput. **21**(5), 792–806 (2017)
10. Chen, Q., Zhang, M., Xue, B.: Structural risk minimization-driven genetic programming for enhancing generalization in symbolic regression. IEEE Trans. Evol. Comput. **23**(4), 703–717 (2018)
11. Deb, K., Gupta, S.: Understanding knee points in bicriteria problems and their implications as preferred solution principles. Eng. Optim. **43**(11), 1175–1204 (2011)
12. Deb, K., Pratap, A., Agarwal, S., Meyarivan, T.: A fast and elitist multiobjective genetic algorithm: NSGA-II. IEEE Trans. Evol. Comput. **6**(2), 182–197 (2002)
13. de França, F.O.: Transformation-interaction-rational representation for symbolic regression: a detailed analysis of SRBench results. ACM Trans. Evol. Learn. (2023)
14. de Franca, F., et al.: Interpretable symbolic regression for data science: analysis of the 2022 competition. arXiv preprint arXiv:2304.01117 (2023)
15. Gaier, A., Ha, D.: Weight agnostic neural networks. In: Advances in Neural Information Processing Systems, vol. 32 (2019)

16. Gonçalves, I., Silva, S.: Balancing learning and overfitting in genetic programming with interleaved sampling of training data. In: Krawiec, K., Moraglio, A., Hu, T., Etaner-Uyar, A.Ş, Hu, B. (eds.) EuroGP 2013. LNCS, vol. 7831, pp. 73–84. Springer, Heidelberg (2013). https://doi.org/10.1007/978-3-642-37207-0_7

17. La Cava, W., Moore, J.H.: Learning feature spaces for regression with genetic programming. Genet. Program Evolvable Mach. **21**, 433–467 (2020)

18. La Cava, W., Silva, S., Danai, K., Spector, L., Vanneschi, L., Moore, J.H.: Multidimensional genetic programming for multiclass classification. Swarm Evol. Comput. **44**, 260–272 (2019)

19. Li, K., Nie, H., Gao, H., Yao, X.: Posterior decision making based on decomposition-driven knee point identification. IEEE Trans. Evol. Comput. **26**(6), 1409–1423 (2021)

20. Muñoz, L., Trujillo, L., Silva, S., Castelli, M., Vanneschi, L.: Evolving multidimensional transformations for symbolic regression with M3GP. Memetic Comput. **11**, 111–126 (2019)

21. Muñoz, M.A., et al.: An instance space analysis of regression problems. ACM Trans. Knowl. Discov. Data (TKDD) **15**(2), 1–25 (2021)

22. Neshatian, K., Zhang, M., Andreae, P.: A filter approach to multiple feature construction for symbolic learning classifiers using genetic programming. IEEE Trans. Evol. Comput. **16**(5), 645–661 (2012)

23. Ni, J., Drieberg, R.H., Rockett, P.I.: The use of an analytic quotient operator in genetic programming. IEEE Trans. Evol. Comput. **17**(1), 146–152 (2012)

24. Ni, J., Rockett, P.: Tikhonov regularization as a complexity measure in multiobjective genetic programming. IEEE Trans. Evol. Comput. **19**(2), 157–166 (2014)

25. Nicolau, M., Agapitos, A.: Choosing function sets with better generalisation performance for symbolic regression models. Genet. Program Evolvable Mach. **22**(1), 73–100 (2021)

26. Olson, R.S., La Cava, W., Orzechowski, P., Urbanowicz, R.J., Moore, J.H.: PMLB: a large benchmark suite for machine learning evaluation and comparison. BioData Min. **10**, 1–13 (2017)

27. Orzechowski, P., La Cava, W., Moore, J.H.: Where are we now? A large benchmark study of recent symbolic regression methods. In: Proceedings of the Genetic and Evolutionary Computation Conference, pp. 1183–1190 (2018)

28. Peng, B., Wan, S., Bi, Y., Xue, B., Zhang, M.: Automatic feature extraction and construction using genetic programming for rotating machinery fault diagnosis. IEEE Trans. Cybern. **51**(10), 4909–4923 (2020)

29. Rachmawati, L., Srinivasan, D.: Multiobjective evolutionary algorithm with controllable focus on the knees of the pareto front. IEEE Trans. Evol. Comput. **13**(4), 810–824 (2009)

30. Ramirez-Atencia, C., Mostaghim, S., Camacho, D.: A knee point based evolutionary multi-objective optimization for mission planning problems. In: Proceedings of the Genetic and Evolutionary Computation Conference, pp. 1216–1223 (2017)

31. Schütze, O., Laumanns, M., Coello, C.A.C.: Approximating the knee of an MOP with stochastic search algorithms. In: Rudolph, G., Jansen, T., Beume, N., Lucas, S., Poloni, C. (eds.) PPSN 2008. LNCS, vol. 5199, pp. 795–804. Springer, Heidelberg (2008). https://doi.org/10.1007/978-3-540-87700-4_79

32. Telikani, A., Tahmassebi, A., Banzhaf, W., Gandomi, A.H.: Evolutionary machine learning: a survey. ACM Comput. Surv. (CSUR) **54**(8), 1–35 (2021)

33. Vanneschi, L., Castelli, M.: Soft target and functional complexity reduction: a hybrid regularization method for genetic programming. Expert Syst. Appl. **177**, 114929 (2021)

34. Vanneschi, L., Castelli, M., Silva, S.: Measuring bloat, overfitting and functional complexity in genetic programming. In: Proceedings of the 12th Annual Conference on Genetic and Evolutionary Computation, pp. 877–884 (2010)

35. Virgolin, M., Alderliesten, T., Bosman, P.A.: On explaining machine learning models by evolving crucial and compact features. Swarm Evol. Comput. **53**, 100640 (2020)

36. Zhang, B.T., Muhlenbein, H., et al.: Evolving optimal neural networks using genetic algorithms with occam's razor. Complex Syst. **7**(3), 199–220 (1993)

37. Zhang, F., Mei, Y., Nguyen, S., Zhang, M.: Evolving scheduling heuristics via genetic programming with feature selection in dynamic flexible job-shop scheduling. IEEE Trans. Cybern. **51**(4), 1797–1811 (2020)

38. Zhang, F., Mei, Y., Nguyen, S., Zhang, M.: Collaborative multifidelity-based surrogate models for genetic programming in dynamic flexible job shop scheduling. IEEE Trans. Cybern. **52**(8), 8142–8156 (2021)

39. Zhang, H., Chen, Q., Xue, B., Banzhaf, W., Zhang, M.: Modular multi-tree genetic programming for evolutionary feature construction for regression. IEEE Trans. Evol. Comput. (2023)

40. Zhang, H., Chen, Q., Xue, B., Banzhaf, W., Zhang, M.: A semantic-based hoist mutation operator for evolutionary feature construction in regression. IEEE Trans. Evol. Comput. (2023). https://doi.org/10.1109/TEVC.2023.3331234

41. Zhang, H., Zhou, A., Chen, Q., Xue, B., Zhang, M.: SR-Forest: a genetic programming based heterogeneous ensemble learning method. IEEE Trans. Evol. Comput. (2023)

42. Zhang, H., Zhou, A., Qian, H., Zhang, H.: PS-tree: a piecewise symbolic regression tree. Swarm Evol. Comput. **71**, 101061 (2022)

43. Zhang, H., Zhou, A., Zhang, H.: An evolutionary forest for regression. IEEE Trans. Evol. Comput. **26**(4), 735–749 (2021)

44. Zhang, X., Tian, Y., Jin, Y.: A knee point-driven evolutionary algorithm for many-objective optimization. IEEE Trans. Evol. Comput. **19**(6), 761–776 (2014)

Short Presentations

Look into the Mirror: Evolving Self-dual Bent Boolean Functions

Claude Carlet[1,2], Marko Durasevic[3], Domagoj Jakobovic[3(✉)], Luca Mariot[4], and Stjepan Picek[1,5]

[1] Department of Mathematics, Université Paris 8, 2 Rue de la Liberté, 93526 Saint-Denis Cedex, France
[2] University of Bergen, Bergen, Norway
[3] University of Zagreb Faculty of Electrical Engineering and Computing, Zagreb, Croatia
{marko.durasevic,domagoj.jakobovic}@fer.hr
[4] Semantics, Cybersecurity & Services Group, University of Twente, Drienerlolaan 5, 7522 Enschede, NB, The Netherlands
l.mariot@utwente.nl
[5] Digital Security Group, Radboud University, PO Box 9010, Nijmegen, The Netherlands
stjepan.picek@ru.nl

Abstract. Bent Boolean functions are important objects in cryptography and coding theory, and there are several general approaches for constructing such functions. Metaheuristics proved to be a strong choice as they can provide many bent functions, even when the size of the Boolean function is large (e.g., more than 20 inputs). While bent Boolean functions represent only a small part of all Boolean functions, there are several subclasses of bent functions providing specific properties and challenges. One of the more interesting subclasses comprises (anti-)self-dual bent Boolean functions.

This paper provides a detailed experimentation with evolutionary algorithms with the goal of evolving (anti-)self-dual bent Boolean functions. We experiment with two encodings and two fitness functions to evolve self-dual bent Boolean functions. Our experiments consider Boolean functions with sizes of up to 16 inputs, and we successfully construct self-dual bent functions for each dimension. Moreover, we notice that the difficulty of evolving self-dual bent functions is similar to evolving bent Boolean functions, despite self-dual bent functions being much rarer.

Keywords: Boolean functions · bent · self-dual bent · evolutionary algorithms

1 Introduction

Bent Boolean functions are interesting mathematical objects. For instance, the codewords of Kerdock codes [10] (which, being provably optimal codes, are important in coding theory) are built from bent functions, and bent functions

M. Giacobini et al. (Eds.): EuroGP 2024, LNCS 14631, pp. 161–175, 2024.
https://doi.org/10.1007/978-3-031-56957-9_10

are used to build so-called bent function sequences for telecommunications [19]. Next, they are related to Golay complementary sequences and bent vectorial functions allow the construction of good codes as well. Bent Boolean functions are also often considered in cryptography since they achieve the highest possible nonlinearity values [17]. Naturally, since bent Boolean functions are not balanced, they cannot be used directly in cryptographic algorithms but can be used to build highly nonlinear Boolean functions [6]. For instance, modified bent Boolean functions can also be used in block ciphers to create S-boxes, as in the case of the CAST-128 and CAST-256 ciphers [1].

Bent Boolean functions have been an important and active research domain for almost 50 years [2,3,16,26]. As such, many works consider how to construct bent Boolean functions. The first and most established approach is to use algebraic constructions. When considering algebraic constructions, dividing them into primary and secondary constructions is common. Primary constructions construct new functions from scratch by leveraging other types of mathematical objects such as permutations and partial spreads [4,15]. Secondary constructions define new functions by starting from existing ones as building blocks [2,3,16,26]. A second approach is to perform numerical simulations. There, one would commonly resort to random search or a certain kind of metaheuristics [5]. Each of those approaches has specific advantages and drawbacks.

Metaheuristics represent an interesting approach for constructing bent Boolean functions as they provide many different function instances and can work for different Boolean function sizes. Unfortunately, when going to larger sizes of Boolean functions, problems arise as there are 2^{2^n} Boolean functions of n variables, and depending on the solution encoding, finding bent Boolean functions can become prohibitively difficult. For instance, one needs 2^n bits to encode a Boolean function of n variables under a bitstring encoding. On the other hand, these difficulties also lead to considering the construction of bent Boolean functions as a benchmark problem [22]. Consequently, we reach the situation where constructing bent Boolean functions becomes interesting for both application-driven reasons and benchmarking.

The interest of the research community is evident from the plethora of works published every year on bent Boolean functions, see, e.g., [2,3,16]. However, we then reach an interesting problem. The research community made significant progress in constructing bent Boolean functions, and the problem can hardly be considered difficult anymore. Indeed, Hrbacek and Dvorak used Cartesian Genetic Programming with various parallelization techniques to evolve bent Boolean functions up to 16 variables [7]. Picek and Jakobovic, on the other hand, used genetic programming to evolve algebraic constructions of bent Boolean functions [20]. Both approaches can provide bent functions of many variables.

In this paper, we focus on a specific subclass called (anti-)self-dual bent Boolean functions. A bent function is called self-dual if equal to its dual, and anti-self-dual if equal to the complement of its dual. We define dual and complement in Sect. 2.2. Such functions are much rarer than bent Boolean functions, satisfying the future requirements for benchmarking. Additionally, while the class of bent Boolean functions is small compared to the class of all Boolean functions, it is still large enough to make enumeration and classification impossible when $n \geq 10$ [2]. Thus, it makes sense to look for subclasses that are more constrained to generate and classify, satisfying also the requirements for practical relevance. We note that the concept of self-dual bent functions is not new and has been discussed already in the 70s [26], when Rothaus observed that "many" bent functions are equal to their duals, i.e., they are self-dual bent functions.

1.1 Related Work

To the best of our knowledge, no works consider metaheuristics for the design of (anti-)self-dual bent Boolean functions. On the other hand, many works focus on generic bent functions. Hrbacek and Dvorak used CGP to evolve bent Boolean functions up to 16 inputs [7]. The authors investigated various configurations of algorithms to speed up the evolution process and succeeded in finding bent functions for sizes between 6 and 16 inputs. Mariot and Leporati used a genetic algorithm to evolve semi-bent Boolean functions by spectral inversion [13]. The novelty of the approach is in using the Walsh-Hadamard spectrum as the genotype instead of the usual truth table-based bitstring encoding. Picek and Jakobovic used GP to evolve algebraic constructions that are then used to construct bent Boolean functions [20]. The authors presented results of up to 24 variables. On a similar research line, Mariot et al. [14] used evolutionary strategies to evolve a secondary construction based on cellular automata for quadratic bent functions.

Husa and Dobai used linear genetic programming to evolve bent Boolean functions [8]. The authors reported better results than related works, and they managed to evolve bent Boolean functions of up to 24 inputs. Picek, Sisejkovic, and Jakobovic investigated immunological algorithms to construct either bent or balanced, highly nonlinear Boolean functions [24]. Picek et al. considered evolving quaternary bent Boolean functions, which are a generalization of bent (binary) Boolean functions [23]. Mariot et al. used evolutionary algorithms to evolve hyper-bent Boolean functions [12]. Hyper-bent Boolean functions are a subclass of bent Boolean functions that also achieve maximum distance from all bijective monomial functions. Interestingly, the authors did not find such functions when using evolutionary algorithms.

We note that there is a large corpus of works considering metaheuristic techniques to construct balanced, highly nonlinear Boolean functions, see, e.g., [21,27] but those works fall outside of our scope since duality is a concept that can be defined for bent Boolean functions only.

1.2 Contributions

This work focuses on the problem of evolving self-dual bent Boolean functions and conducts an extensive experimental evaluation. Our main contributions are:

– To the best of our knowledge, we are the first ones considering metaheuristics to evolve (anti-)self-dual bent Boolean functions. More precisely, we consider

evolutionary algorithms and conduct experiments with two solution representations and two fitness functions. Our experiments with 8 to 16 inputs show
we can successfully evolve such functions with the tree encoding.
- While (anti-)self-dual bent Boolean functions are rarer than general bent
  Boolean functions, we show that the difficulty for evolutionary algorithms is
  rather similar for both problems. Interestingly, when evolving for the nonlinearity property only, better results are obtained for anti-self-dual bent functions, while one would intuitively expect the same difficulty for self-dual and
  anti-self-dual bent functions.
- Surprisingly, when evolving (anti)-self-dual bent functions in 16 variables,
  we managed to find such functions, while the corresponding search for the
  general bent functions gave no successful results in our experiments. This
  could indicate that for a large number of variables, constraining the search
  to a subclass can even be beneficial for metaheuristics.

2   Preliminaries

This section covers all necessary background information related to bent Boolean
functions used throughout the paper. We start with some basic algebraic notation
and then move to basic representations and properties of Boolean functions.

2.1   Notation

We denote by $\mathbb{F}_2$ the finite field with two elements, where sum and multiplication
correspond respectively to the XOR (denoted as $\oplus$) and logical AND (denoted
by concatenation) of the two operands. Given a positive integer $n \in \mathbb{N}^+$, the
n-dimensional vector space over $\mathbb{F}_2$ is denoted as $\mathbb{F}_2^n$, while $\mathbb{F}_{2^n}$ is the finite field
with 2^n elements.

Up to isomorphism, there exists a unique field $\mathbb{F}_{2^n}$ of order 2^n for all $n \in \mathbb{N}$.
Since this field is also an n-dimensional vector space, we can endow the vector
space $\mathbb{F}_2^n$ with the structure of the field $\mathbb{F}_2^n$ when convenient. The usual inner
product of $a, b \in \mathbb{F}_2^n$ equals $a \cdot b = \bigoplus_{i=1}^n a_i b_i$.

A Boolean function is any mapping $f : \mathbb{F}_2^n \to \mathbb{F}_2$ from $\mathbb{F}_2^n$ to $\mathbb{F}_2$, and it can be
uniquely represented by its truth table, which is the list of pairs of function inputs
(in $\mathbb{F}_2^n$) and function values. The value vector is the binary vector composed of
all $f(x), x \in \mathbb{F}_2^n$, where some total order has been fixed on $\mathbb{F}_2^n$ (most commonly,
the lexicographic order). Since the size of the value vector equals 2^n, the number
of Boolean functions of n variables is 2^{2^n}, i.e., they grow super-exponentially in
n. In practice, exhaustive enumeration of the set of n-variable Boolean functions
becomes unfeasible already for $n > 5$ (see Table 1).

The Walsh-Hadamard transform W_f is another unique representation of a
Boolean function that measures the correlation between $f(x)$ and the linear
functions $a \cdot x$ (with the sum being calculated in $\mathbb{Z}$):

$$W_f(a) = \sum_{x \in \mathbb{F}_2^n} (-1)^{f(x) \oplus a \cdot x}. \tag{1}$$

Table 1. The number of Boolean functions with a given number of variables.

n	3	4	5	6	7	8	9	10	11	12
2^{2^n}	2^8	2^{16}	2^{32}	2^{64}	2^{128}	2^{256}	2^{512}	2^{1024}	2^{2048}	2^{4096}

The Walsh-Hadamard transform is very useful in cryptography, as many properties relevant to attacks on stream and block cipher models can be evaluated through it. To compute the Walsh-Hadamard transform efficiently, one can use the fast Walsh-Hadamard transform [2].

2.2 Definitions, Properties, and Bounds

A Boolean function f is balanced if it takes the output value 1 exactly the same number 2^{n-1} of times as value 0 when the input ranges over $\mathbb{F}_2^n$.

The minimum Hamming distance between a Boolean function f and all affine functions, i.e., the functions of algebraic degree at most 1 (of the same number of variables as f), is called the nonlinearity of f. The nonlinearity can be expressed in terms of the Walsh-Hadamard coefficients of f as follows [2]:

$$nl_f = 2^{n-1} - \frac{1}{2} \max_{a \in \mathbb{F}_2^n} |W_f(a)|. \tag{2}$$

The Parseval relation states that the sum of the squared Walsh-Hadamard values is constant for any Boolean function $f : \mathbb{F}_2^n \to \mathbb{F}_2$:

$$\sum_{a \in \mathbb{F}_2^n} W_f(a)^2 = 2^{2n}. \tag{3}$$

This relation implies that the arithmetic mean of $W_f(a)^2$ equals 2^n. Since the maximum of $W_f^2(a)$ is equal to or larger than its arithmetic mean, we can deduce that $\max_{a \in \mathbb{F}_2^n} |W_f(a)|$ must be equal to or larger than $2^{\frac{n}{2}}$.

This implies that for every n-variable Boolean function, f satisfies the so-called covering radius bound:

$$nl_f \leq 2^{n-1} - 2^{\frac{n}{2}-1}. \tag{4}$$

Boolean functions can satisfy the bound in Eq. (3) with equality if and only if $W_f(a) = \pm 2^{\frac{n}{2}}$ for all $a \in \mathbb{F}_2^n$. Such functions are also called *bent*, and they reach the maximum possible nonlinearity value $2^{n-1} - 2^{n/2-1}$. Remark that bent Boolean functions exist only for n even since the nonlinearity is an integer.

For a bent function f on $\mathbb{F}_{2^n}$, we define its *dual* as the Boolean function $\tilde{f} : \mathbb{F}_{2^n} \to \mathbb{F}_2$ satisfying:

$$2^{\frac{n}{2}} (-1)^{\tilde{f}(x)} = W_f(x) \text{ for all } x \in \mathbb{F}_{2^n}. \tag{5}$$

Table 2. Number of bent and self-dual bent Boolean functions.

n	2	4	6	8
#bent	8	896	5425430528	$2^{106.3}$
#self-dual	2	20	42896	104960

The dual $\widetilde{f}$ of a bent function is also bent. A bent function f is said to be self-dual if $\widetilde{f}(x) \oplus f(x) = 0$ for all $x \in \mathbb{F}_2^n$, and anti-self-dual if $\widetilde{f}(x) \oplus f(x) = 1$. Stated differently, a bent function is called self-dual if it is equal to its dual and anti-self-dual if it is equal to the complement of its dual.

Bent Boolean functions are rare, and we know the exact numbers of bent Boolean functions for $n \leq 8$ only. Self-dual bent functions are even rarer. We provide the numbers of bent and self-dual bent functions in Table 2. Note that for the self-dual bent functions of 8 variables, we have results for quadratic functions (those with the algebraic degree at most two) only. The total number of self-dual bent functions is thus larger. There are as many anti-self-dual bent functions as there are self-dual bent functions.

For further information about bent Boolean functions and their properties, we refer interested readers to [2,3,11,16].

3   Experimental Setup

In this section, we discuss the solution representations, fitness functions, and evolutionary algorithm parameters.

3.1   Solution Representations

Bitstring Encoding. When evolving Boolean functions, the most common approach to encoding a solution is the bitstring representation, as discussed in [5]. The bitstring represents the truth table of the function upon which the algorithm operates directly. For a Boolean function of n variables, the truth table is encoded as a bitstring of length 2^n. With this representation, the corresponding variation operators we use are the simple bit mutation, which inverts a randomly selected bit, and a mixing mutation, which shuffles the bits within a randomly selected substring. For the crossover operators, we use the one-point crossover, which combines a new solution, taking the first portion from one and the second from the other parent, with a randomly selected breakpoint. The second operator is the uniform crossover, which randomly selects a bit to be copied from both parents at each position into the child bitstring with uniform probability. Each time a crossover or mutation operation is invoked by the evolutionary algorithm, a random operator is chosen among the described ones. We decided to use multiple mutation and crossover operators as our preliminary analysis showed this gives better results. It would be interesting to study in more detail the influence of various operators or their combinations on the results.

Tree Encoding. The second approach in our experiments uses tree-based GP to represent a Boolean function in its symbolic form. GP and its variants (such as Cartesian Genetic Programming [18]) have already been extensively used in the evolution of Boolean functions and have been able to produce human-competitive results [5,21]. In this case, we represent a candidate solution by a tree whose leaf nodes correspond to the input variables $x_1, \cdots, x_n \in \mathbb{F}_2$. The internal nodes are Boolean operators that combine the inputs received from their children and propagate their output to the respective parent nodes.

Our experiments use the NOT function that takes a single argument, the function set operating on two arguments: OR, XOR, AND, AND2[1], and XNOR, and the function IF, which takes three arguments and returns the second one if the first one evaluates to true and the third one otherwise. The output of the root node is the output value of the Boolean function. The corresponding truth table of the function $f : \mathbb{F}_2^n \to \mathbb{F}_2$ is determined by evaluating the tree over all possible 2^n assignments of the input variables at the leaf nodes. Each GP individual is evaluated according to the truth table it produces. The genetic operators used for GP are simple tree, uniform, size fair, one-point, and context preserving crossover [25] (selected at random each time crossover is performed), and subtree mutation. Again, the choice of using multiple genetic operators was done based on the preliminary experiments.

Since the search size grows rapidly with the number of inputs, we expect the bitstring encoding to perform much worse than the GP encoding, which is in accordance with most of the previous works, as discussed before. However, we include both encodings for completeness and a more reliable estimate of the problem's difficulty.

3.2   Fitness Functions

To evolve bent Boolean functions, one only needs to check that the maximal absolute value in the Walsh-Hadamard transform equals $2^{\frac{n}{2}}$ (see Eq. (2)). For self-dual functions, each Walsh-Hadamard coefficient must not only be equal to this absolute value: considering Eq. (5), its sign must agree with the corresponding output value in the function's truth table. For instance, if $f(a) = 0$ for $a \in \mathbb{F}_2^n$, the corresponding coefficient $W_f(a)$ in the Walsh-Hadamard transform must assume the value of $+2^{\frac{n}{2}}$, and $-2^{\frac{n}{2}}$ otherwise; for anti-self-dual functions, the previous values are inverted.

The remark above suggests the following strategy for our first fitness function (denoted as fit_1): count the number of entries in the Walsh-Hadamard transform whose absolute value is equal to $2^{\frac{n}{2}}$ and, at the same time, the sign of the value matches the corresponding output value in the truth table. Formally, given $f : \mathbb{F}_2^n \to \mathbb{F}_2$, its fitness score under fit_1 is defined as:

$$fit_1(f) = |\{a \in \mathbb{F}_2^n : W_f(a) = 2^{\frac{n}{2}} \cdot (-1)^{f(a)}\}| . \tag{6}$$

---

[1] The function AND2 behaves the same as the function AND but with the second input inverted.

Since the number of entries in the Walsh-Hadamard transform is equal to the truth table size (2^n), the range of this fitness function is $[0, \ldots, 2^n]$, where 2^n denotes the optimal value that corresponds to a self-dual bent function. Note that this fitness function will drive the search towards the bent and self-dual criteria at the same time, as opposed to a lexicographic approach that would first try to achieve a bent function and then additionally optimize duality.

The second fitness function we employ takes a closer look into the deviation of each Walsh-Hadamard entry from the desired value. Apart from the number of correct values, as evaluated by fit_1, we sum the absolute differences (from either $2^{\frac{n}{2}}$ or $-2^{\frac{n}{2}}$) of every incorrect coefficient, and divide the sum with the product of the maximal possible difference $(2^{\frac{n}{2}})$ by the total number of entries (2^n). Consequently, the deviation part is normalized in $[0, 1]$, and its difference from 1 is simply added to the number of correct entries computed through fit_1. Hence, the fitness score of $f : \mathbb{F}_2^n \to \mathbb{F}_2$ under fit_2 is formally defined as:

$$fit_2(f) = fit_1(f) + \left[1 - \frac{\sum_{a \in \mathbb{F}_2^n} \left| 2^{\frac{n}{2}} \cdot (-1)^{f(a)} - W_f(a) \right|}{2^n \cdot 2^{\frac{n}{2}}} \right]. \qquad (7)$$

The integer part of fit_2 always equals the value obtained with fit_1. In particular, when the normalized sum of the deviations is 0 (that is, we reached an optimal solution), the difference from 1 is not added to fit_1. Thus, the optimal fitness value for fit_2 is the same as fit_1, i.e., 2^n.

Finally, since (anti-)self-dual functions are a subset of bent functions, we also included experiments that optimize only the nonlinearity property, trying to obtain bent functions. In this experiment, only the tree-based GP representation is used since this approach has shown to be the most efficient one in evolving bent Boolean functions [5].

4   Experimental Results

In this section, we first describe the settings adopted for our experimental evaluation of the evolutionary algorithms described in the previous section to design (anti)-self-dual bent functions. Then, we report the results of the experiments.

4.1   Experimental Settings

Both bitstring and GP encoding employed the same evolutionary algorithm: a steady-state selection scheme with a 3-tournament elimination operator. In each iteration of the algorithm, three individuals are chosen at random from the population for the tournament, and the worst one in terms of fitness value is eliminated. The two remaining individuals in the tournament are used by the crossover operator to generate a new child individual, which then undergoes mutation with individual mutation probability $p_{mut} = 0.5$. Finally, the mutated child takes the place of the eliminated individual in the population. The implementation relies on the open-source framework ECF.[2]

---

[2] Evolutionary Computation Framework, http://solve.fer.hr/ECF/.

The population size in all experiments was 500, and the termination criteria were set to 10^6 evaluations. Finally, each experiment was repeated 30 times. In order to test whether there is a significant difference between the different tested approaches, the Kruskal–Wallis test is used with the Bonferroni correction method and Dunn's post hoc test. The results are considered to be statistically different if a p–value below 0.05 is obtained. The maximum tree depth for the GP representation was based on a set of preliminary experiments and was set to $\max(5, n - 5)$, where n is the number of Boolean variables (which is also the size of the terminal set).

4.2 Evolving (Anti)-Self-Dual Bent Boolean Functions

The results obtained when evolving (anti)-self-dual bent Boolean functions are outlined in Table 3. The results demonstrate that in all cases when using GP, we can find the target (anti-)self-dual function in at least one run, which is evident from the fact that the best fitness value was equal to 2^n, meaning that every Walsh-Hadamard entry was correct. Table 4 shows the number of successful runs (out of 30) in which GP managed to find (at least one) self-dual or anti-self-dual function.

As expected, the bitstring representation (denoted with TT) obtained inferior results when compared to GP and could not reach the target values, i.e., it did not find a self-dual or anti-self-dual function in even one of the runs (for this reason, we omit the results for anti-self-dual bent functions). When observing the influence of the fitness function variant used in the experiments, we see no difference between the two fitness functions when using the GP representation, as in both cases, the target functions were obtained. On the other hand, for the TT representation, there are some minor differences, although not consistently in favor of a single fitness function.

Since all tested GP variants achieved the optimal value, Table 4 outlines the number of times each variant achieved the optimal value from the 30 executions that were performed. For 6 and 8 variables the self-dual GP variant with fitness 1 obtained the optimal value in all 30 executions, whereas the other variants sometimes obtained sub–optimal values. However, as the number of variables increases, the variants that evolve anti-self-dual functions usually achieve the optimum value in more cases, especially when using fitness fit_2. For the largest number of variables, there is no clear winner. However, the self-dual GP with fit_1 achieves the optimal value the least number of times. This shows that a single algorithm variant does not perform consistently well across all the problem sizes that were considered.

Table 3. Best obtained fitness values when optimizing for self-dual and anti-self-dual bent Boolean functions.

Representation	Fitness function	Variables					
		6	8	10	12	14	16
self-dual TT	fit_1	41	69	101	169	257	471
self-dual TT	fit_2	43	70	103	168	256	481
self-dual GP	fit_1	64	256	1024	4096	16384	65536
self-dual GP	fit_2	64	256	1024	4096	16384	65536
anti-self-dual GP	fit_1	64	256	1024	4096	16384	65536
anti-self-dual GP	fit_2	64	256	1024	4096	16384	65536

Table 4. The number of runs in which the optimal value was obtained.

Representation	Fitness function	Variables					
		6	8	10	12	14	16
self-dual GP	fit_1	30	30	22	14	4	2
self-dual GP	fit_2	29	28	26	13	8	4
anti-self-dual GP	fit_1	28	30	20	11	9	4
anti-self-dual GP	fit_2	30	27	26	17	11	3

To further outline the differences between the different variants, Table 5 provides the average number of generations each GP variant required to obtain the optimal solutions. Until now, we see that the results are not conclusive and that no tested variant always reaches the optimal solution when a different number of variables are used. However, some methods seem to reach optimal solutions faster than others. For the smallest variable sizes, the self-dual GP with fit_2 obtains the optimal results in the least number of generations. This variant is also the fastest when 12 or 14 variables are considered. For 16 variables, it was second best, while the anti-self-dual GP with fit_1 was the best. Therefore, self-dual GP with fit_2 seems to be, on average, the fastest variant when it comes to the number of generations required to obtain the optimal solution. On the other hand, self-dual GP with fit_1 is consistently the slowest when a larger number of variables are considered.

Table 5. The average number of generations required to obtain the optimal solution.

Representation	Fitness function	Variables					
		6	8	10	12	14	16
self-dual GP	fit_1	74	142	300	565	751	1743
self-dual GP	fit_2	46	126	324	296	485	923
anti-self-dual GP	fit_1	91	153	250	310	653	399
anti-self-dual GP	fit_2	55	249	151	401	673	1715

The results are additionally presented as box plots to outline better the distribution obtained across all executions of the GP algorithm (TT is not considered here due to its inferior performance). Figure 1 provides the results obtained by GP for Boolean functions considering 14 variables. The figure shows that fit_2 leads to somewhat better results in both cases as the median values obtained by it are slightly higher. However, Fig. 2 demonstrates that for 16 variables, neither fitness function definition consistently achieved a better performance. Therefore, it is impossible to conclude that either fitness function definition is significantly better. Still, we recommend the second fitness since it provides more information and can potentially direct the search better.

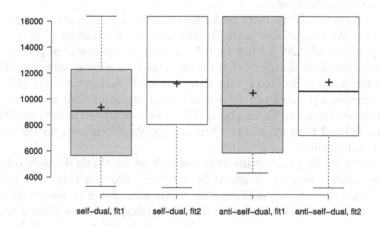

Fig. 1. Box plots of the results obtained when optimizing for self-dual and anti-self-dual bent Boolean functions considering functions of 14 variables.

Figures 3 and 4 show the nonlinearity levels of the obtained Boolean functions considering 14 and 16 variables (denoted on the y–axis), respectively. In this case, we compare the results obtained when optimizing only for nonlinearity, denoted in the figure under the column 'bent', against those obtained by optimizing for self-dual or anti-self-dual functions. First, we can notice from the results that when optimizing for nonlinearity, the results' dispersion is rather small, and the

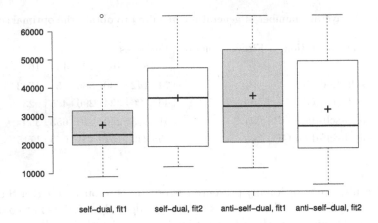

Fig. 2. Box plots of the results obtained when optimizing for self-dual and anti-self-dual bent Boolean functions considering functions of 16 variables.

algorithm obtains similar results across all runs. In the remaining cases, the results are significantly more dispersed, and often, poor nonlinearity levels are obtained. However, such behavior is expected as, in these cases, the nonlinearity was not optimized directly, and therefore, it is expected that functions with poor nonlinearity may be found.

To test whether GP variants performed significantly different from each other, the Kruskal–Wallis test was used. For the number of variables 6, 8, 10, 12, and 14, the p–values of 0.29, 0.13, 0.19, 0.58, 0.17 were obtained, respectively. This means that for these numbers of variables, there were no significant differences between the results obtained by the four tested GP variants. On the other hand, for 16 variables a p–value of 0.04 was obtained, meaning that there is a significant difference between the methods. The post hoc test demonstrated that the difference existed between the two fitness versions when used for the evolution of self–dual functions.

When considering the nonlinearity levels, it seems that fit_2 more often leads to better results, meaning it might be a better choice in this case. Regarding self-dual and anti-self-dual optimization, the results also demonstrate that better nonlinearity values were obtained when optimizing for anti-self-dual functions. However, more experimental runs should be conducted to verify if this effect is statistically significant. An additional interesting observation from the results is that when 16 variables were considered, optimizing for self-dual or anti-self-dual functions always resulted in at least one bent - and at the same time dual - function (with a nonlinearity level equal to 32640), which is evident from Table 3. On the other hand, by optimizing directly for nonlinearity, we did not obtain functions with this nonlinearity value.

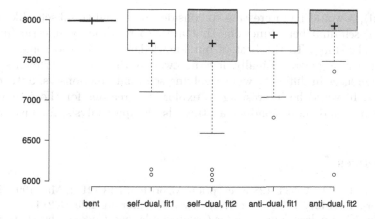

Fig. 3. Achieved nonlinearity levels - 14 variables.

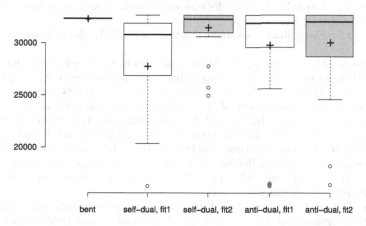

Fig. 4. Achieved nonlinearity levels - 16 variables.

5 Conclusions and Future Work

This paper tackles the problem of evolving (anti)-self-dual bent functions. Our results show that the tree encoding is much more efficient than the bitstring (truth table-based) one. This is aligned with the related works and results on general bent Boolean functions. We can observe that the problem also does not seem much more difficult than evolving general bent functions since GP manages to achieve it for every dimension. We also observe that when evolving for nonlinearity only (i.e., maximizing the nonlinearity value), better results are obtained for anti-self-dual bent functions, which (intuitively) should not be easier than considering self-dual bent functions.

In future work, one could consider evolving constructions of (anti-)self-dual bent functions. Moreover, since the problem of evolving (anti-)self-dual bent functions is not difficult enough (or, more precisely, is still simpler than we expected), it would be interesting to consider even smaller subsets, like if there are (anti)-self-dual bent functions that are also rotation symmetric (invariant under cyclic shift). To the best of our knowledge, this is also something that is not known in general. Finally, it is noteworthy that our experiments showed some differences in difficulty when evolving self-dual functions vs. anti-self-dual functions. It would be interesting to explore the reasons for those differences. One option could be to conduct a fitness landscape analysis, for instance, like in [9].

References

1. Adams, C.: The CAST-128 Encryption Algorithm. RFC 2144, May 1997. https://doi.org/10.17487/RFC2144, https://www.rfc-editor.org/info/rfc2144
2. Carlet, C.: Boolean Functions for Cryptography and Coding Theory. Cambridge University Press, Cambridge (2021). https://doi.org/10.1017/9781108606806
3. Carlet, C., Mesnager, S.: Four decades of research on bent functions. Des. Codes Cryptogr. **78**(1), 5–50 (2016)
4. Dillon, J.F.: Elementary Hadamard difference sets. Ph.D. thesis, Univ. of Maryland (1974)
5. Djurasevic, M., Jakobovic, D., Mariot, L., Picek, S.: A survey of metaheuristic algorithms for the design of cryptographic Boolean functions. Cryptogr. Commun. **15**(6), 1171–1197 (2023). https://doi.org/10.1007/s12095-023-00662-2
6. Dobbertin, H.: Construction of bent functions and balanced Boolean functions with high nonlinearity. In: Preneel, B. (ed.) FSE 1994. LNCS, vol. 1008, pp. 61–74. Springer, Heidelberg (1995). https://doi.org/10.1007/3-540-60590-8_5
7. Hrbacek, R., Dvorak, V.: Bent function synthesis by means of cartesian genetic programming. In: Bartz-Beielstein, T., Branke, J., Filipič, B., Smith, J. (eds.) PPSN 2014. LNCS, vol. 8672, pp. 414–423. Springer, Cham (2014). https://doi.org/10.1007/978-3-319-10762-2_41
8. Husa, J., Dobai, R.: Designing bent Boolean functions with parallelized linear genetic programming. In: Proceedings of the Genetic and Evolutionary Computation Conference Companion, pp. 1825–1832. GECCO '17, Association for Computing Machinery, New York, NY, USA (2017). https://doi.org/10.1145/3067695.3084220
9. Jakobovic, D., Picek, S., Martins, M.S., Wagner, M.: Toward more efficient heuristic construction of Boolean functions. Appl. Soft Comput. **107**, 107327 (2021). https://doi.org/10.1016/j.asoc.2021.107327, https://www.sciencedirect.com/science/article/pii/S1568494621002507
10. Kerdock, A.: A class of low-rate nonlinear binary codes. Inf. Control **20**(2), 182–187 (1972)
11. MacWilliams, F.J., Sloane, N.J.A.: The Theory of Error-Correcting Codes. Elsevier, Amsterdam, North Holland (1977). ISBN: 978-0-444-85193-2
12. Mariot, L., Jakobovic, D., Leporati, A., Picek, S.: Hyper-bent Boolean functions and evolutionary algorithms. In: Sekanina, L., Hu, T., Lourenço, N., Richter, H., García-Sánchez, P. (eds.) EuroGP 2019. LNCS, vol. 11451, pp. 262–277. Springer, Cham (2019). https://doi.org/10.1007/978-3-030-16670-0_17
13. Mariot, L., Leporati, A.: A genetic algorithm for evolving plateaued cryptographic Boolean functions. In: Dediu, A.-H., Magdalena, L., Martín-Vide, C. (eds.) TPNC 2015. LNCS, vol. 9477, pp. 33–45. Springer, Cham (2015). https://doi.org/10.1007/978-3-319-26841-5_3

14. Mariot, L., Saletta, M., Leporati, A., Manzoni, L.: Heuristic search of (semi-)bent functions based on cellular automata. Nat. Comput. **21**(3), 377–391 (2022)
15. McFarland, R.L.: A family of difference sets in non-cyclic groups. J. Comb. Theory Ser. A **15**(1), 1–10 (1973). https://doi.org/10.1016/0097-3165(73)90031-9, https://www.sciencedirect.com/science/article/pii/0097316573900319
16. Mesnager, S.: Bent Functions. Springer International Publishing, Cham (2016). https://doi.org/10.1007/978-3-319-32595-8
17. Mesnager, S.: Linear codes from functions. In: Huffman, W.C., Solé, J.L.K.P. (eds.) A Concise Encyclopedia of Coding Theory. p. 94 pages in Chapter 20. Press/Taylor and Francis Group (2021)
18. Miller, J.F.: An empirical study of the efficiency of learning Boolean functions using a cartesian genetic programming approach. In: Proceedings of the 1st Annual Conference on Genetic and Evolutionary Computation, vol. 2, pp. 1135–1142. GECCO'99, Morgan Kaufmann Publishers Inc., San Francisco, CA, USA (1999)
19. Olsen, J., Scholtz, R., Welch, L.: Bent-function sequences. IEEE Trans. Inf. Theory **28**(6), 858–864 (1982)
20. Picek, S., Jakobovic, D.: Evolving algebraic constructions for designing bent Boolean functions. In: Proceedings of the Genetic and Evolutionary Computation Conference 2016, pp. 781–788. GECCO '16, Association for Computing Machinery, New York, NY, USA (2016). https://doi.org/10.1145/2908812.2908915
21. Picek, S., Jakobovic, D.: Evolutionary computation and machine learning in security. In: Proceedings of the Genetic and Evolutionary Computation Conference Companion, pp. 1572–1601. GECCO '22, Association for Computing Machinery, New York, NY, USA (2022). https://doi.org/10.1145/3520304.3534087
22. Picek, S., Jakobovic, D., O'Reilly, U.M.: Cryptobench: benchmarking evolutionary algorithms with cryptographic problems. In: Proceedings of the Genetic and Evolutionary Computation Conference Companion, pp. 1597–1604. GECCO '17, Association for Computing Machinery, New York, NY, USA (2017). https://doi.org/10.1145/3067695.3082535
23. Picek, S., Knezevic, K., Mariot, L., Jakobovic, D., Leporati, A.: Evolving bent quaternary functions. In: 2018 IEEE Congress on Evolutionary Computation, CEC 2018, Rio de Janeiro, Brazil, 8–13 July 2018, pp. 1–8. IEEE (2018)
24. Picek, S., Sisejkovic, D., Jakobovic, D.: Immunological algorithms paradigm for construction of Boolean functions with good cryptographic properties. Eng. Appl. Artif. Intell. **62**, 320–330 (2017). https://doi.org/10.1016/j.engappai.2016.11.002, http://www.sciencedirect.com/science/article/pii/S0952197616302044
25. Poli, R., Langdon, W.B., McPhee, N.F.: A Field Guide to Genetic Programming. Lulu Enterprises Ltd., UK (2008)
26. Rothaus, O.: On bent functions. J. Comb. Theory Ser. A **20**(3), 300–305 (1976)
27. Yan, L., et al.: IGA: an improved genetic algorithm to construct weightwise (almost) perfectly balanced Boolean functions with high weightwise nonlinearity. In: Proceedings of the 2023 ACM Asia Conference on Computer and Communications Security, pp. 638–648. ASIA CCS '23, Association for Computing Machinery, New York, NY, USA (2023). https://doi.org/10.1145/3579856.3590337

An Algorithm Based on Grammatical Evolution for Discovering SHACL Constraints

Rémi Felin(✉) , Pierre Monnin , Catherine Faron ,
and Andrea G. B. Tettamanzi

Université Côte d'Azur, Inria, CNRS, I3S, Sophia-Antipolis, Nice, France
{Remi.Felin,Pierre.Monnin,Catherine.Faron,andrea.tettamanzi}@inria.fr

Abstract. The continuous evolution of heterogeneous RDF data has led to an increase of inconsistencies on the Web of data (i.e. missing data and errors) that we assume to be inherent to RDF data graphs. To improve their quality, the W3C recommendation SHACL allows to express various constraints that RDF data must conform to and detect nodes violating them. However, acquiring representative and meaningful SHACL constraints from complex and very large RDF data graphs is very challenging and tedious. Consequently, several recent works focus on the automatic generation of these constraints. We propose an approach based on grammatical evolution (GE) for extracting representative SHACL constraints by mining an RDF data graph. This approach uses a probabilistic SHACL validation framework to consider the inherent errors in RDF data. The results highlight the relevance of this approach in discovering SHACL shapes inspired by association rule patterns from a real-world RDF data graph.

Keywords: Grammatical Evolution · Shape Mining · Web of Data

1 Introduction

Over the years, the Web has witnessed an increasing publication of heterogeneous data graphs forming the "Web of Data", aimed at being consumed by humans and artificial agents. These graphs are represented using the RDF (Resource Description Format) standard and queried using the SPARQL standard. RDF relies on a triple model: an RDF triple $< s, p, o >$ expresses a relation p between a subject s and an object o. An **RDF data graph** is a set of interconnected RDF triples which terms are IRIs, literals and blank nodes (anonymous resources): s is a resource (a IRI or a blank node), p is a IRI, and o is any RDF term. RDF data can be serialized with several syntaxes among which Turtle is the simplest and most readable one. It uses qualified names with *prefixes* associated to namespaces to simplify the notation of IRIs, e.g. `foaf:name` is parsed into <http://xmlns.com/foaf/0.1/name>. The increase of available and heterogeneous RDF data is

M. Giacobini et al. (Eds.): EuroGP 2024, LNCS 14631, pp. 176–191, 2024.
https://doi.org/10.1007/978-3-031-56957-9_11

the result of global initiatives for producing open data, e.g. Linked Open Data[1]. It is well known that this evolution has revealed issues of errors and incompleteness in real-world RDF data graphs and we consider it essential to recognize the principle that these inconsistencies are inherent to RDF data graphs.

To improve the quality of RDF data, the W3C recommended SHACL [11] as a standard to represent constraints on RDF data graphs. **SHACL shapes** are instances of sh:NodeShape[2] that allow targeting a specific set of nodes in an RDF data graph and assessing them against a set of SHACL constraints, i.e. searching possible nodes in the RDF data graph that do not conform to the shape. Overall, the evaluation process considers a shapes graph (i.e., a set of SHACL shapes) and evaluate them against an RDF graph. SHACL thus addresses the requirements for RDF data quality control, contributing to reducing the inherent inconsistencies in RDF data graphs.

We are interested in the extraction of SHACL shapes that express domain constraints from RDF data graphs. Given that SHACL is a relatively new language, real-world data graphs have a minimal set of associated SHACL shapes. This issue, extensively discussed by Rabbati et al. [28], has led the community to explore various research directions. We can distinguish between approaches aiming to extract SHACL shapes from an ontology (i.e., the "schema" associated with RDF graphs) [4,27], approaches aiming to extract SHACL shapes by mining regularities from RDF facts [3,8,25], and approaches combining both [1]. In general, one of the most significant challenges in extracting SHACL shapes is scaling these methods to handle substantial data graphs. Only Fernandez-Álvarez et al. [8] address this problem by optimizing machine memory consumption. Ontology-based approaches are limited to the coverage degree of the ontologies regarding the RDF data graph, which impacts the type of constraints that can be extracted. Some approaches for extracting SHACL shapes from RDF data take errors and incompleteness into account, using a threshold of tolerance [8] or quality measures [25], but they have limitations regarding the types of SHACL shapes they can extract. For instance, the method proposed in [25] is limited to the extraction of a specific type of rules that can be later translated into SHACL shapes. Boneva et al. [3] propose a semi-automatic method to discover SHACL shapes (via a user interface), while Pandit et al. [27] suggest a manual construction of SHACL shapes using Ontology Design Patterns.

We aim to address the limitations regarding the kind of SHACL shapes that can be extracted from an RDF data graph, which motivated the following research question: *How to automatically discover SHACL shapes from RDF data?* We believe that a generative approach for the automatic construction of SHACL candidate shapes using RDF data is one of those that can achieve this ideal. To this end, we propose a mining method based on **Grammatical Evolution (GE)**, which is a particular type of genetic programming. In this approach, it is feasible to automatically generate variable-length expressions in any language [26] using well-defined grammars, composed of production rules.

---

[1] https://lod-cloud.net/.

[2] Or <http://www.w3.org/ns/shacl#NodeShape>.

Grammatical Evolution has been the subject of ongoing work in recent years [30]. Recent work discusses some limitations concerning the grammar design [6], their complexity and a "poor" initialisation of individuals [9]. These limitations have been the subject of contributions intended to propose general guidelines for grammar design [23], automated techniques for finding optimal parameters [2], or new techniques for improving the population initialisation [24]. However, the two major and recurring problems with this method are *redundancy* between individuals and low *locality* [13,16], i.e. "how well neighbouring genotypes correspond to neighbouring phenotypes" [29]. Lourenço et al. have proposed an extension to this approach called Structured Grammatical Evolution (SGE) [12,14,15], which enables one-to-one mapping between genes and non-terminals belonging to the grammar, with effective responses to these two problems. Other approaches aim to introduce a probabilistic approach which is based on a probabilistic selection of production rules during the genotype building phase [10,17]. Mégane et al. extend their approach with a probabilistic SGE [18,19] to improve their results and resolve these two issues. In these works, some well-known benchmarks are used to assess the effectiveness of the proposed models, like *Santa Fe Trail* and *Boston Housing*, but their application to tasks related to RDF data mining has not been demonstrated yet. Only Nguyen and Tettamanzi have proposed an adaptation of GE for extracting OWL disjointness axioms [21] and complex disjointness axioms [22], with some promising results.

In this paper, we propose an algorithm based on Grammatical Evolution for generating candidate SHACL shapes using a BNF grammar and RDF data as input. The algorithm exploits a probabilistic framework for SHACL validation, which is required given the heterogeneity and incompleteness inherent in open RDF data. The approach is validated by applying it to the extraction of SHACL shapes from a set of RDF data relating to the scientific domain. The remainder of the paper is organized as follows: in Sect. 2, we present the design of BNF grammars describing candidate SHACL shapes; in Sect. 3, we present the probabilistic SHACL validation (Sect. 3.1) and how we use it to define an acceptance measure and a fitness function (Sect. 3.2) in order to evaluate candidate shapes; in Sect. 4 we present a recombination operator responding to the redundancy problem and variation operators used to ensure a broad exploration of the solution space; in Sect. 5, we present the experiments carried out on a real-world RDF dataset. The results of our experiments are presented in Sect. 5.2, and we conclude with a discussion of future research in Sect. 6.

2   BNF Grammars of SHACL Shapes

In order to produce and exploit well-formed SHACL shapes as individuals in an evolutionary process, we defined a BNF grammar compliant with the SHACL W3C recommendation [11]. Figure 1 presents a subset of this BNF grammar to produce shapes targeting nodes of a specified class (`sh:targetClass`) and constraining them to be linked through the predicate `rdf:type` to another specified class. It should be noted that the two classes may be the same. The grammar provides the phenotypic and genotypic characterisation for each individual using a set of *static rules* and *dynamic rules*. The dynamic rules system, proposed by Nguyen and Tettamanzi [21,22], allows the mapping between rules and RDF data. They wrote BNF grammars to build candidate OWL axioms and exploit them with an evolutionary algorithm based on GE. The static rules are the *immutable* components of the phenotypic character whereas the dynamic rules are the *problem instance-dependent* components of the phenotypic character, where each rule has one or many possible values depending on the considered RDF dataset, and each value is identified by a genotype. However, their dynamic rules were hard-coded in their system. In contrast, we extended the dynamic rules design by directly enabling the user to write embedded SPARQL queries as value of one or more production rules in the grammar in order to perform the mapping with the desired granularity.

```
1  <Shape>      := "a " <NodeShape>
2  <NodeShape>  := "sh:NodeShape; " <ShapeBody>
3  <ShapeBody>  := "sh:targetClass " <Class> "; " <ShapeProp>
4  <ShapeProp>  := "sh:property [ " + <PropBody> " ] ."
5  <PropBody>   := "sh:path rdf:type; sh:hasValue " <Class> " ;"
6  <Class>      := "SPARQL ?x rdf:type ?Class"
```

Fig. 1. An extract of the BNF grammar for SHACL shapes

To illustrate, in Fig. 1, the <Class> non-terminal (used in lines 3 and 5) is defined by a dynamic rule to extract all possible classes from the RDF dataset using a SPARQL query. The keyword SPARQL in line 6 is used to specify the query graph pattern to be matched on RDF data; the result of this query is the set of nodes in the RDF data graph $\mathcal{C}$ bound to variable ?Class in the query graph pattern: $\mathcal{C} = \{c_i, i \in [1, n]\}$. The final step is to replace the initial value of the rule <Class>, i.e. the SPARQL query "SPARQL ?x a ?Class", by the SPARQL results $\mathcal{C}$, i.e. $c_1 \mid c_2 \mid \cdots \mid c_n$. The whole process is presented in Fig. 2.

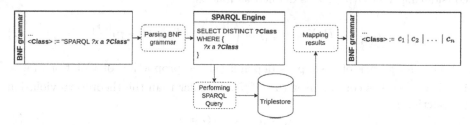

Fig. 2. Dynamic rules process based on the BNF grammar presented in Fig. 1

Using the BNF grammar presented in Fig. 1, the genotype of an individual is a pair of codons $[i, j]$, which are decoded into two classes from the dataset using a classic genotype-phenotype mapping, and produce the following phenotype structure:

```
"a sh:NodeShape ; sh:targetClass c_i ; sh:property [ sh:path rdf:type ;
                    sh:hasValue c_j" ; ] ."
```

It is noteworthy that the proposed grammar can be extended to produce a wider array of SHACL shapes using a variable-length template, e.g. replacing the rule <ShapeProp> from Fig. 1 by <ShapeProp> := <Prop> <ShapeProp> | <Prop> where <Prop> := "sh:property [" + <PropBody> "] .". Such an extended grammar would produce SHACL shapes specifying one or more constraints (depending on the chosen length).

3   Probabilistic SHACL Validation as a Fitness Function

3.1   Preliminaries

We rely on the probabilistic framework for SHACL validation proposed by Felin et al. [7] to assess SHACL shapes considering inherent inconsistencies from RDF data. It extends the standard evaluation of RDF data against SHACL shapes by considering a *physiological* error proportion p in the RDF data, i.e. a possible acceptable violation rate. *Physiological* errors in RDF data graph are inherent errors whose origins can be diverse, e.g. from collaborative building of large RDF data graphs (e.g. Wikidata) or automatically constructed RDF data graphs (e.g. DBpedia). In the rest of this section, we summarize the principles of this model.

Considering an RDF data graph v, the *support* of a shape s, v_s, is the set of RDF triples in v targeted by s (and therefore tested during the validation). The *reference cardinality* (refCard) of s is the cardinality of its support: $|v_s|$. The *confirmations* and *violations* of s, respectively v_s^+ and v_s^-, are the sets of triples that, respectively, are consistent with s and violate s: $v_s^+ \cap v_s^- = \emptyset$, and $v_s = v_s^+ \cup v_s^-$. The probabilistic model for SHACL validation relies on a binomial distribution $X \sim B(|v_s|, p)$ where p is the physiological error proportion. The *likelihood* to observe a number of violations $|v_s^-|$ in the support of a shape s considering $X \sim B(|v_s|, p)$ is defined as follows:

$$L_{|v_s^-|} = P(X = |v_s^-|) = \binom{|v_s|}{|v_s^-|} \cdot p^{|v_s^-|} \cdot (1-p)^{|v_s^+|} \tag{1}$$

The *acceptance* of a shape s depends on the proportion of violations for s, i.e. $\hat{p}_s = \frac{|v_s^-|}{|v_s|}$: s is consistent with v if $\hat{p}_s$ is smaller than the theoretical violation proportion p:

$$\hat{p}_s \leq p \implies KG \models s \tag{2}$$

In the case where $\hat{p}_s > p$, a Chi-Square Goodness of Fit test is performed using the *test statistic* X_s^2 defined as follows:

$$X_s^2 = \sum_{i=1}^{k} \frac{(n_i - T_i)^2}{T_i} \sim \chi_{k-1;\alpha}^2 \tag{3}$$

where k is the total number of groups, i.e. $k = 2$, n_i is the observed number of individuals and T_i is the theoretical number of individuals. The acceptance of the null hypothesis H_0 implies the acceptance of s:

$$X_s^2 \leq \chi_{k-1;\alpha}^2 \implies KG \models s \qquad (4)$$

3.2 Acceptability and Fitness Score

In order to iteratively produce a final population of SHACL shapes expressing some domain constraints that are implicit in an RDF dataset, we propose a fitness function based on an acceptability measure of a shape combined with the probabilistic framework presented.

The Acceptability of a SHACL shape s, $A(s)$, regarding an RDF dataset, depends on the error rate when validating this RDF dataset against s, which depends on the theoretical error proportion p. The acceptability measure $A(s) \in [0, 1]$ of a SHACL shape is defined by:

$$A(s) = \begin{cases} 1 & \text{if } \hat{p}_s \leq p \text{ or } X_s^2 \leq \chi_{k-1;\alpha}^2 \quad \text{(Equation (2) and 4)}, \\ \frac{L_{|v_s^-|}}{P(X=|v_s|\times p)} & \text{otherwise} \quad \text{(Equation (1))}. \end{cases} \qquad (5)$$

In the computation of $A(s)$, for the hypothesis testing, we consider a margin error $\alpha = 0.05$. Therefore, the *critical value* is $\chi_{k-1;0.05}^2 = 3.84$. When $A(s) = 1$, s is acceptable and is (probably) selected as one of the most fit individuals, in the sense that it captures some domain knowledge extracted from the RDF data.

In the case where the null hypothesis is rejected, $A(s) \neq 1$ and so s is not acceptable but it may be considered in the grammatical evolution algorithm for crossover or mutation operations. For this purpose, $A(s)$ is equal to the likelihood of s normalized by the maximal value of the probability mass function for a binomial distribution $X \sim B(|v_s|, p)$. It ensures a better distribution of $A(s)$ values between 0 and 1 in contrast to the lonely likelihood value $L_{|v_s^-|}$ and therefore avoids excessively penalising individuals who are "close" to being acceptable but for whom the likelihood is very low.

The Fitness Function of a SHACL shape s, $F(s)$, regarding an RDF dataset, combines its acceptability $A(s)$ and the cardinality of its confirmations $|v_s^+|$:

$$\forall s \in P, \quad F(s) = |v_s^+| \times A(s) \qquad (6)$$

4   Variation and Recombination Operators

In this paper, we adapt the main components of the GE variation operators to discover SHACL shapes over RDF data, considering the problem of *redundancy* and *low locality* presented as the main issues of GE by the community.

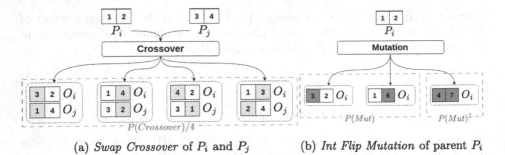

(a) *Swap Crossover of P_i and P_j*   (b) *Int Flip Mutation of parent P_i*

Fig. 3. Representation of GE operators and their probabilities of occurrence

The *redundancy* is observed when many genotypes map the same phenotype expression [13]. Based on this fact, we adapt the recombination phase to filter every offspring using **a phenotypic comparison**: Algorithm 1 presents the recombination of selected individuals $\mathcal{S}$ among the whole population $\mathcal{P}$ ($\mathcal{E}$ represents the elite individuals). Line 11 describes the conditions for integrating an offspring i into the replacement population $\mathcal{R}$: i is integrated into $\mathcal{R}$ if the phenotypic expression of i is not already observed among the elitist individuals $\mathcal{E}$ and the replacement individuals $\mathcal{R}$. As a consequence, we avoid the reflection of the redundancy in the final population: $\forall i \in \mathcal{P}, \nexists j \in \mathcal{P} : i \equiv j$.

Algorithm 1: Recombination of a population $\mathcal{P}$

    Data: elite individuals $\mathcal{E}$ **and** selected individuals $\mathcal{S}$
    Result: replacement population $\mathcal{R}$

1 $\mathcal{R} \leftarrow \{\}$;
2 **while** $|\mathcal{R}| \neq |\mathcal{P}| - |\mathcal{E}|$ **do**
3     $\mathcal{C} \leftarrow \{\}$;
4     $p_1 \leftarrow \mathcal{S}[random() * |\mathcal{S}|]$;
5     $p_2 \leftarrow \mathcal{S}[random() * |\mathcal{S}|]$; $\triangleright$ *random() * $|\mathcal{S}|$ as integer;*
6   **if** $p_1 \neq p_2$ **then**
7        $\mathcal{C} \leftarrow p_1 \cup p_2$;
8        $\mathcal{C} \leftarrow crossover(\mathcal{C})$; $\triangleright$ Figure 3a;
9        $\mathcal{C} \leftarrow mutation(\mathcal{C})$; $\triangleright$ Figure 3b;
10   **for** $i \in \mathcal{C}$ **do**
11   **if** $i \notin \mathcal{E} \cup \mathcal{R}$ **and** $|\mathcal{R}| \neq |\mathcal{P}| - |\mathcal{E}|$ $\triangleright$ **Phenotypic** comparison
12   **then**
13           $\mathcal{R} \leftarrow \mathcal{R} \cup i$;
14   **end**
15 **end**
16 **return** $\mathcal{R}$

In this context, the *locality* issue is dependent on the neighbourhood between the selected rules from parent to offspring. Consequently, some results from the variation operators presented in Fig. 3 can lead to a low locality, i.e. a very different offspring, but also to a fairly strong locality. Considering the grammar presented in Fig. 1, a modification of the first codon (impacting the value of the sh:targetClass c_i) significantly changes the meaning of the phenotypic trait, as SHACL validation is performed on the nodes instantiating c_i: a new production rule c'_i replacing c_i leads to a locality as low as the proximity (e.g. common instances) between them. The modification of the last codon has a lower impact on SHACL validation since the targeted nodes are the same, resulting in a fairly strong locality (even if the meaning of the phenotype is different).

5 Experiences

5.1 Shape Mining over the Covid-on-the-Web Dataset

Setup. *Covid-on-the-Web*[3] [20] is an RDF dataset produced from the *COVID-19 Open Research Dataset (CORD-19)*. It describes articles and named entities identified in these articles and linked to *Wikidata* entities. We consider a subset containing 18.79% of the articles and 0.01% of the named entities. This dataset contains 226,647 RDF triples, 20,912 distinct articles and 6,331 distinct named entities.

We consider the mining of SHACL shapes representing **association rules** between *Wikidata entities*[4], i.e., rules of the form $\mathcal{X} \rightarrow \mathcal{Y}$. We use the BNF grammar presented in Fig. 1 to generate candidate shapes. Each candidate involves a first Wikidata entity, i.e., the *antecedent* called $\mathcal{X}$, and targets nodes n (i.e., scientific articles) typed by this entity using the sh:targetClass property. The proposed constraint verifies if these nodes are also typed by a second Wikidata entity, i.e., the *consequent* $\mathcal{Y}$, using the sh:hasValue constraint applied on the rdf:type property.

Concerning the aim of these experiments, we considered the diversity and the discovery of a large set of acceptable shapes as the most important aspects. While it is obvious that a resource-intensive parameter setting (high population size, high effort, etc.) would provide the best results at the cost of a long computation time, we have focused on a reasonable set of parameters in order to minimize the computation time invested.

We used an implementation of the presented algorithm combined with the probabilistic SHACL validation engine implemented in the *Corese* semantic Web factory [5]. We considered a theoretical error proportion $p = 0.5$ (i.e., *physiological* error), according to the experiment's results on this dataset reported by Felin et al. [7]: this value p maximises the mean value of the likelihood measure L (see Eq. (1)).

---

[3] https://github.com/Wimmics/CovidOnTheWeb.
[4] https://www.wikidata.org/wiki/Wikidata:Main_Page.

The experiments have been performed on a server equipped with an Intel(R) Xeon(R) CPU E5-2637 v2 processor at 3.50 GHz clock speed, with 172 GB of RAM, 1 TB of disk space running under the Ubuntu 20.06.4 LTS 64-bit operating system.

Recall. In order to assess the ability of our approach to find acceptable candidates in the solution space of our problem, we define the recall R of our algorithm. The recall provides the rate of distinct solutions found by our algorithm among the total number of solutions $\mathcal{A}$. Let Ω be the set of all possible (acceptable and non-acceptable) pairs $(\mathcal{X}, \mathcal{Y})$ of distinct named entities extracted from the *Covid-on-the-web dataset*. We may compute its cardinality using the number of distinct entities in the dataset: $|\Omega| = 6,331 \times 6,330 = \mathbf{40,075,230}$.

To estimate the number of acceptable shapes in Ω, we sample a random subset $\Omega' \subseteq \Omega$, representative of the solution space. To ensure the representativity of Ω', we determine its minimal size using the *Cochran* formula: $|\Omega'| = \frac{z^2 \times p \times (1-p)}{m^2} = \frac{2.58^2 * 0.5^2}{0.02^2} \approx \mathbf{4,161}$, where z is the standard normal z-table with a confidence level of 99% (so $z \approx 2.58$), m is the tolerated margin of error (2%); p' the probability that the candidate shape is acceptable (unknown in this context so $p' = 0.5$).

Consequently, we generated 4,161 distinct and random shapes which have been evaluated over the *Covid-on-the-Web* subgraph with probabilistic SHACL validation: the results show that only 2 shapes in Ω' are accepted, *i.e.*, 0.05% of the total. This allows us to estimate the total number of acceptable shapes $\mathcal{A}$: $|\mathcal{A}| = |\Omega| \times 0.0005 = \mathbf{20,037.6}$.

Finally, the recall of our algorithm $R(x)$, *i.e.*, how well the x acceptable shapes cover the solution space, is defined by:

$$\mathbf{R(x)} = \frac{x}{|\mathcal{A}|} \times 100 \tag{7}$$

5.2   Results

$|\mathcal{P}|/E$ **Choice.** We assessed our approach with manually defined small population sizes ($|\mathcal{P}|$) and quite low *effort* values E and we analysed the effects of the ratio $|\mathcal{P}|/E$. This corresponds to verifying if our algorithm can find credible and surprising candidate shapes using a minimum investment of CPU time. Consequently, we performed 10 executions of our algorithm using the different parameter settings presented in Table 1 (90 in total) and analyzed the final whole population $\mathcal{P}$ and the final elitist subset $\mathcal{E}$ ($\mathcal{E} \subseteq \mathcal{P}$). Each configuration has been assessed regarding the following metrics: the average fitness value $\overline{\mathbf{F}}$; the average rate of accepted shapes $\overline{\%\mathbf{A}}$; the average likelihood value $\overline{\mathbf{L}}$; the average CPU time (in ms) for evaluating an individual $\overline{\mathbf{T}}$ and the average recall $\overline{\mathbf{R}}$.

According to the results presented in Table 2, a gradual increase of the effort E tends to enhance the global quality of candidate shapes into $\mathcal{P}$: all the metrics related to the individual quality ($\overline{F}$, $\overline{\%A}$, $\overline{L}$ and $\overline{R}$) are significantly better

Table 1. Used parameters to analyse the impact of $|\mathcal{P}|/E$ choice.

Parameters	Value(s)
GE	
$|P|$	$\{100;\ 200;\ 500\}$
Effort (E)	$\{5,000;\ 10,000;\ 20,000\}$
% Selection ($\mathcal{E}$)	20%
% Selection ($\mathcal{R}$)	40%
Selection type	Tournament
% Tournament	25%
Crossover type - P	Swap (Fig. 3a) - 75%
Mutation type - P	Int Flip (Fig. 3b) - 5%
Probabilistic SHACL	
Confidence level α	5%
Theoretical inconsistencies proportion p	50%

Table 2. Results obtained using the parameters presented in Table 1. Best result for each metric is in bold and second best underlined. Highlighted columns are the best.

		$|P| = 100$			$|P| = 200$			$|P| = 500$	
		E = 5,000	E = 10,000	E = 20,000	E = 5,000	E = 10,000	E = 20,000	E = 5,000	E = 10,000	E = 20,000
From $\mathcal{P}$	$\overline{F}$	0.79 ± 1.4	0.64 ± 1.05	**1.52 ± 1.7**	0.53 ± 0.81	1.24 ± 1.83	<u>1.51 ± 1.14</u>	0.9 ± 0.97	1.07 ± 1	1.3 ± 0.87
	%A	1.9 ± 2.77	4 ± 2.71	**10.1 ± 4.18**	2 ± 0.75	3.3 ± 2.15	<u>8.05 ± 3.11</u>	1.56 ± 0.76	2.36 ± 1.05	4.98 ± 2.33
	$\overline{L}$	2.55 ± 1.77	4.86 ± 1.66	**6.74 ± 2.07**	2.42 ± 0.97	4.41 ± 1.85	<u>6.18 ± 1.13</u>	1.25 ± 0.29	2.07 ± 0.59	4.47 ± 1.29
	$\overline{T}$	18 ± 2.73	<u>17.98 ± 2.96</u>	21.94 ± 3.73	**16.84 ± 2.4**	19.22 ± 3.03	21.14 ± 3.65	19.21 ± 2.28	18.39 ± 3.71	19.31 ± 3.71
	$\overline{R}$	0.01 ± 0.01	0.02 ± 0.01	0.05 ± 0.02	0.02 ± 0.01	0.03 ± 0.02	<u>0.08 ± 0.03</u>	0.04 ± 0.02	0.06 ± 0.03	**0.13 ± 0.06**
From $\mathcal{E}$	$\overline{F}$	3.91 ± 7	2.77 ± 5.28	**7.41 ± 8.48**	2.65 ± 4.06	6.11 ± 9.15	**7.41 ± 5.74**	4.41 ± 4.9	5.29 ± 4.99	<u>6.36 ± 4.39</u>
	%A	9.5 ± 13.83	18.5 ± 12.92	**49.5 ± 19.64**	9.5 ± 4.22	15.75 ± 11.31	<u>39.5 ± 15.27</u>	6.7 ± 3.13	11.2 ± 4.94	24.1 ± 10.99
	$\overline{L}$	9.36 ± 6.74	16.18 ± 4.47	**21.51 ± 4.88**	8.21 ± 3.23	14.96 ± 5.84	<u>20.73 ± 2.25</u>	4.03 ± 1.1	7.22 ± 1.86	15.19 ± 3.67
	$\overline{T}$	10.6 ± 2.45	9.35 ± 1.43	7.68 ± 1.26	9.22 ± 1.28	7.53 ± 1.11	**6.65 ± 0.65**	10.64 ± 2.83	7.81 ± 0.94	<u>7.21 ± 2.66</u>
	$\overline{R}$	0.01 ± 0.01	0.02 ± 0.01	0.05 ± 0.02	0.02 ± 0.01	0.03 ± 0.02	<u>0.08 ± 0.03</u>	0.03 ± 0.02	0.06 ± 0.02	**0.12 ± 0.06**

regardless of the population size $|P|$. This is clearer regarding the elitist part of the population $\mathcal{E}$.

Globally, it appears that the smallest $|\mathcal{P}|$ with a high effort provide the best results: the results obtained with ($|\mathcal{P}| = 100$; $E = 20,000$) and ($|\mathcal{P}| = 200$; $E = 20,000$) are very similar, except the proportion of acceptable shapes in $\mathcal{E}$ (respectively 49.5% and 39.5%). For each metric, we performed a *Mann-Withney-Wilconox* (Table 3) test to highlight any differences between each obtained result: it appears that only the average recall $\overline{R}$ values (from $\mathcal{P}$ and $\mathcal{E}$) are significantly different between them (<0.05) which suggests that the choice ($|\mathcal{P}| = 200$; $E = 20,000$) is the best one for this measure.

Selection ($\mathcal{R}$) Pressure. We assume that the smallest population combined with the highest effort, *i.e.*, ($|\mathcal{P}| = 100$; $E = 20,000$), is the best choice for analysing selective pressure and learning about its impact on metrics. Conse-

Table 3. *Mann-Whitney-Wilcoxon* test: comparison between the results obtained for $(|P| = 100; E = 20,000)$ and $(|P| = 200; E = 20,000)$ with $\alpha = 0.05$.

From $\mathcal{P}$		From $\mathcal{E}$	
Metrics	P-value	Metrics	P-value
$\overline{F}$	0.528	$\overline{F}$	0.529
$\overline{\%A}$	0.198	$\overline{\%A}$	0.210
$\overline{L}$	0.684	$\overline{L}$	0.796
$\overline{T}$	0.631	$\overline{T}$	0.076
$\overline{R}$	**0.037**	$\overline{R}$	**0.028**

quently, we have studied different selection types: *Scaled Roulette Wheel* and *Tournament* with the different settings presented in Table 4.

Table 4. Parameters used to analyse the impact of the selective pressure on $\mathcal{R}$.

Parameters	Value(s)
GE	
$|P|$	100
Effort (E)	20,000
% Selection ($\mathcal{E}$)	20%
Selection type	{Scaled Roulette Wheel; Tournament}
% Selection ($\mathcal{R}$)	{20%; 40%; 60%}
% Tournament	{10%; 25%; 50%}
Crossover type - P	Swap (Fig. 3a) - 75%
Mutation type - P	Int Flip (Fig. 3b) - 5%
Probabilistic SHACL	
Confidence level α	5%
Theoretical inconsistencies proportion p	50%

The results obtained with the *Scaled Roulette Wheel* selection are presented in Table 5. They highlight that the metrics are enhanced with a high selection rate ($S = 60\%$) even though no very good candidates have been found. It appears that a high selection rate enhances the exploration of the solution space, resulting in a relatively strong difference between the results observed for $\mathcal{P}$ and the elitist subset $\mathcal{E}$.

The results obtained with the *Tournament* selection are presented in Table 6 and we identify the same trend. For $S = 20\%$, the global difference of results between $\mathcal{P}$ and $\mathcal{E}$ is quite low, enhancing the homogeneity of the population $\mathcal{P}$ whereas a high selection rate ($S = 60\%$) reflects the heterogeneity of the population because the global difference of results between $\mathcal{P}$ and $\mathcal{E}$ is high. We

Table 5. Results obtained using the *Scaled Roulette Wheel* selection and parameters presented in Table 4: best result for each metric is in bold and second best underlined.

		$S = 20\%$	$S = 40\%$	$S = 60\%$
From $\mathcal{P}$	$\overline{F}$	0.86 ± 1.06	$\mathbf{1.78 \pm 2.99}$	0.56 ± 0.16
	$\overline{\%A}$	8.7 ± 3.83	7.3 ± 5.33	$\mathbf{11.7 \pm 3.06}$
	$\overline{L}$	7.85 ± 2.46	$\mathbf{8.85 \pm 1.89}$	$\underline{8.25 \pm 2.6}$
	$\overline{T}$	20.25 ± 6.26	$\underline{19.86 \pm 8.61}$	$\mathbf{19.83 \pm 5.78}$
	$\overline{R}$	0.04 ± 0.02	0.04 ± 0.03	$\mathbf{0.06 \pm 0.02}$
From $\mathcal{E}$	$\overline{F}$	4.27 ± 5.29	$\mathbf{8.87 \pm 14.97}$	2.74 ± 0.8
	$\overline{\%A}$	$\underline{43.5 \pm 19.16}$	36 ± 25.47	$\mathbf{58 \pm 14.94}$
	$\overline{L}$	$\underline{24.33 \pm 4.36}$	$\mathbf{25.25 \pm 5.5}$	23.94 ± 4.31
	$\overline{T}$	8.49 ± 1.29	$\underline{7.77 \pm 1.46}$	$\mathbf{7.38 \pm 1.19}$
	$\overline{R}$	$\underline{0.04 \pm 0.02}$	0.04 ± 0.03	$\mathbf{0.06 \pm 0.02}$

can see from the difference between the average time for the whole population vs the average time for the elite that a high selection rate S tends to make the population vary considerably while maintaining a very good elite population which ensures shapes of good quality in the elite with a wide exploration in the global population, favouring the discovery of heterogeneous and potentially interesting shapes.

Table 6. Results obtained using the *Tournament* selection and parameters presented in Table 4: best result for each metric is in bold, second best underlined. The highlighted column corresponds to the *reference* results presented in Table 2.

		$S = 20\%$			$S = 40\%$			$S = 60\%$		
		$Tour = 10\%$	$Tour = 25\%$	$Tour = 50\%$	$Tour = 10\%$	$Tour = 25\%$	$Tour = 50\%$	$Tour = 10\%$	$Tour = 25\%$	$Tour = 50\%$
From $\mathcal{P}$	$\overline{F}$	1.78 ± 1.93	$\underline{1.82 \pm 2.44}$	1.08 ± 1.98	1.2 ± 1.54	1.52 ± 1.7	1.71 ± 2.42	0.99 ± 0.83	$\mathbf{2.32 \pm 2}$	1.78 ± 2.33
	$\overline{\%A}$	$\mathbf{11.1 \pm 5.2}$	8.9 ± 5.45	8.1 ± 5.43	9.9 ± 4.51	10.1 ± 4.18	9.7 ± 4.27	8.3 ± 3.97	$\underline{10.7 \pm 4.99}$	8.4 ± 3.27
	$\overline{L}$	5.55 ± 2.6	6.84 ± 2.81	6.98 ± 2.58	6.11 ± 1.14	6.74 ± 2.07	6.03 ± 2.02	6.12 ± 1.32	6.36 ± 2.1	$\mathbf{7.1 \pm 1.32}$
	$\overline{T}$	22.92 ± 9.12	$\mathbf{16.4 \pm 3.4}$	$\underline{18.88 \pm 6.41}$	23.93 ± 5.93	21.94 ± 3.73	21.66 ± 7.37	28.44 ± 7.5	27.89 ± 6.46	27.9 ± 6.62
	$\overline{R}$	$\mathbf{0.06 \pm 0.03}$	0.04 ± 0.03	0.04 ± 0.03	$\underline{0.05 \pm 0.02}$	$\underline{0.05 \pm 0.02}$	0.05 ± 0.02	0.04 ± 0.02	$\underline{0.05 \pm 0.03}$	0.04 ± 0.02
From $\mathcal{E}$	$\overline{F}$	8.83 ± 9.64	$\underline{9.02 \pm 12.21}$	5.34 ± 9.94	5.93 ± 7.69	7.41 ± 8.48	8.47 ± 12.13	4.9 ± 4.14	$\mathbf{11.55 \pm 10.02}$	8.86 ± 11.68
	$\overline{\%A}$	$\mathbf{55.5 \pm 25.98}$	43.5 ± 26.46	39.5 ± 25.65	49 ± 21.96	49.5 ± 19.64	48.5 ± 21.35	41 ± 19.12	$\underline{52.5 \pm 24.41}$	42 ± 16.36
	$\overline{L}$	18.96 ± 6.49	19.1 ± 3.73	21.6 ± 4.95	21.22 ± 2.5	21.51 ± 4.88	19.77 ± 3.69	22.68 ± 4.24	$\underline{24.25 \pm 5.67}$	$\mathbf{24.46 \pm 3.09}$
	$\overline{T}$	7.54 ± 1	8.09 ± 1.07	7.76 ± 0.79	7.4 ± 1.52	7.68 ± 1.26	7.25 ± 1.51	$\mathbf{6.93 \pm 0.43}$	$\underline{7.06 \pm 1.18}$	8.02 ± 1.38
	$\overline{R}$	$\mathbf{0.06 \pm 0.03}$	0.04 ± 0.03	0.04 ± 0.03	$\underline{0.05 \pm 0.02}$	$\underline{0.05 \pm 0.02}$	0.05 ± 0.02	0.04 ± 0.02	$\underline{0.05 \pm 0.02}$	0.04 ± 0.02

Acceptable Shapes. Among all the conducted experiments, we have discovered a set of 1,766 distinct and acceptable shapes. An overview of these results is provided in Table 7. Some of these shapes have been accepted with a very high violation rate ($>50\%$) but can be easily validated, e.g. the candidate implying the following rule (`gene expression profiling` → `gene expression`) is easily understandable and acceptable even if it implies 52.6% of violations. Moreover, 46.38% of the whole has been accepted after performing hypothesis testing which shows it has a valuable impact on the acceptance of shapes and so on the mining process.

However, some of these shapes require final validation from experts due to the complexity, e.g. the following rule (`chemokine` → `cytokine`) has been automatically accepted and validated after some research: "*Chemokines [...] are a family of small cytokines*"[5]. Rule (`tlr9` → `toll-like receptor`) has been accepted but must be validated by experts.

Table 7. Overview of the distinct and acceptable shapes discovered from all the performed experiments.

Metrics	$\overline{F}$	$\overline{L}$	$\overline{T}$	$\overline{R}$
Values	19.49	19.14	7.93	8.91

The discovery of very good candidates from some of these experiments impacts the standard deviation of many values $\overline{F}$ and $\overline{\%A}$ (some of these are higher than the mean value). We also note some trivial shapes with identical classes for the `sh:targetClass` and the constraint `sh:hasValue` which implies a perfect acceptance of these candidates, *i.e.*, without any violations, and so a very good fitness value. These can be generated because of production rules selection (with a modulo operator) and a *quasi-infinite* range for codon definition. However, this is a fairly rare occurrence: we observe it among (only) 132 candidates from the 1,766 acceptable shapes, *i.e.*, 7.47%. We suggest accepting a low occurrence of these shapes being discovered (even if they are uninteresting) in order to avoid any negative impact on the exploration of the solution space.

The $\overline{T}$ value presented in Table 7 and the correlation between the number of violations and the CPU time presented in Fig. 4 suggest that the CPU time required to invest in an evolutionary process is maximal at the beginning, then decreases as the average number of violations decreases (and therefore when the shapes become more and more acceptable). Considering this expected evolution and that the average time is low, this demonstrates the relevance of this evolutive approach to the discovery of SHACL shapes over an RDF data graph and appears to be suitable for scalability.

---

[5] https://en.wikipedia.org/wiki/Chemokine.

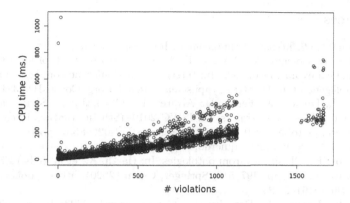

Fig. 4. CPU time spent for the SHACL validation of each discovered shape considering the number of violations

6 Conclusion

In this paper, we have proposed a framework using a grammatical evolution method to extract candidate SHACL shapes from an RDF dataset based on a manually defined BNF grammar. The proposed algorithm provides an effective response to the redundancy problem whereas the low locality appears to be *problem-dependent*. Additionally, the generative evolution allows to tackle the requirement of a broad exploration of the huge search space of possible SHACL shapes to discover acceptable ones. The framework uses a probabilistic SHACL validation process with an acceptability measure and a fitness function to evaluate candidate forms and retain the best ones. Experiments show that our approach captures interesting SHACL shapes describing domain constraints from a real-world RDF dataset. In addition, it provides an effective way to discover a large set of heterogeneous shapes in RDF data, weighting the errors that these may imply and adapting the mining of these shapes. Our future research will focus on studying the scalability of this approach to larger RDF datasets such as DBpedia ($> 50M$ triples). In addition, we plan to extend the approach to the exploration of complex shapes, e.g. shapes with multiple constraints. Finally, it appears important to study the application (and adaptation) of various algorithms proposed by the community, e.g. Structured Grammatical Evolution [15], to the SHACL shapes mining.

Supplemental Material Statement. The source code, RDF datasets and obtained results are available in a public repository.[6]

Acknowledgements. This work has been partially founded by the 3IA Côte d'Azur "Investments in the Future" project managed by the National Research Agency (ANR) with the reference number ANR-19-P3IA-0002.

---

[6] https://github.com/RemiFELIN/RDFMining/tree/eurogp_2024.

References

1. shaclgen 0.2.5.2. https://github.com/uwlib-cams/shaclgen
2. Ali, M.S., Kshirsagar, M., Naredo, E., Ryan, C.: Autoge: a tool for estimation of grammatical evolution models. In: International Conference on Agents and Artificial Intelligence (2021). https://api.semanticscholar.org/CorpusID:232106265
3. Boneva, I., Dusart, J., Fernández Alvarez, D., Gayo, J.E.L.: Shape designer for ShEx and SHACL constraints. In: ISWC 2019–18th International Semantic Web Conference, October 2019. https://hal.science/hal-02268667, poster
4. Cimmino, A., Fernández-Izquierdo, A., García-Castro, R.: Astrea: automatic generation of SHACL shapes from ontologies. In: Harth, A., et al. (eds.) ESWC 2020. LNCS, vol. 12123, pp. 497–513. Springer, Cham (2020). https://doi.org/10.1007/978-3-030-49461-2_29
5. Cérés, R., Corby, O., Demairy, E.: Corese, March 2023. https://github.com/Wimmics/corese
6. Dick, G., Whigham, P.A.: Initialisation and grammar design in grammar-guided evolutionary computation (2022)
7. Felin, R., Faron, C., Tettamanzi, A.G.B.: A framework to include and exploit probabilistic information in SHACL validation reports. In: ESWC (2023)
8. Fernandez-Álvarez, D., Labra-Gayo, J.E., Gayo-Avello, D.: Automatic extraction of shapes using shexer. Knowl.-Based Syst. **238**, 107975 (2022). https://doi.org/10.1016/j.knosys.2021.107975
9. Harper, R.: GE, explosive grammars and the lasting legacy of bad initialisation. In: IEEE Congress on Evolutionary Computation, pp. 1–8 (2010). https://doi.org/10.1109/CEC.2010.5586336
10. Kim, H.-T., Ahn, C.W.: A new grammatical evolution based on probabilistic context-free grammar. In: Handa, H., Ishibuchi, H., Ong, Y.-S., Tan, K.-C. (eds.) Proceedings of the 18th Asia Pacific Symposium on Intelligent and Evolutionary Systems - Volume 2. PALO, vol. 2, pp. 1–12. Springer, Cham (2015). https://doi.org/10.1007/978-3-319-13356-0_1
11. Kontokostas, D., Knublauch, H.: Shapes constraint language (SHACL). W3C recommendation, W3C (2017)
12. Lourenço, N., Assunção, F., Pereira, F., Costa, E., Machado, P.: Structured grammatical evolution: a dynamic approach, pp. 137–161 (2018). https://doi.org/10.1007/978-3-319-78717-6_6
13. Lourenço, N., Ferrer, J., Pereira, F.B., Costa, E.: A comparative study of different grammar-based genetic programming approaches. In: McDermott, J., Castelli, M., Sekanina, L., Haasdijk, E., García-Sánchez, P. (eds.) EuroGP 2017. LNCS, vol. 10196, pp. 311–325. Springer, Cham (2017). https://doi.org/10.1007/978-3-319-55696-3_20
14. Lourenço, N., Pereira, F., Costa, E.: Unveiling the properties of structured grammatical evolution. Genet. Program. Evolvable Mach. **17** (2016). https://doi.org/10.1007/s10710-015-9262-4
15. Lourenço, N., Pereira, F.B., Costa, E.: SGE: a structured representation for grammatical evolution. In: Bonnevay, S., Legrand, P., Monmarché, N., Lutton, E., Schoenauer, M. (eds.) EA 2015. LNCS, vol. 9554, pp. 136–148. Springer, Cham (2016). https://doi.org/10.1007/978-3-319-31471-6_11
16. Medvet, E.: A comparative analysis of dynamic locality and redundancy in grammatical evolution. In: McDermott, J., Castelli, M., Sekanina, L., Haasdijk, E., García-Sánchez, P. (eds.) EuroGP 2017. LNCS, vol. 10196, pp. 326–342. Springer, Cham (2017). https://doi.org/10.1007/978-3-319-55696-3_21

17. Mégane, J., Lourenço, N., Machado, P.: Probabilistic grammatical evolution (2021)
18. Mégane, J., Lourenço, N., Machado, P.: Co-evolutionary probabilistic structured grammatical evolution. In: Proceedings of the Genetic and Evolutionary Computation Conference. ACM, July 2022. https://doi.org/10.1145/3512290.3528833
19. Mégane, J., Lourenço, N., Machado, P.: Probabilistic structured grammatical evolution. In: 2022 IEEE Congress on Evolutionary Computation (CEC). IEEE, July 2022. https://doi.org/10.1109/cec55065.2022.9870397
20. Michel, F., et al.: COVID-on-the-web: knowledge graph and services to advance COVID-19 research. In: Pan, J.Z., et al. (eds.) The Semantic Web – ISWC 2020. ISWC 2020. LNCS, vol. 12507, pp. 294–310. Springer, Cham (2020). https://doi.org/10.1007/978-3-030-62466-8_19
21. Nguyen, T.H., Tettamanzi, A.G.B.: An evolutionary approach to class disjointness axiom discovery. In: Barnaghi, P.M., Gottlob, G., Manolopoulos, Y., Tzouramanis, T., Vakali, A. (eds.) WI 2019 - IEEE/WIC/ACM International Conference on Web Intelligence, pp. 68–75. ACM, Thessaloniki, Greece, October 2019. https://doi.org/10.1145/3350546.3352502
22. Nguyen, T.H., Tettamanzi, A.G.B.: Grammatical evolution to mine OWL disjointness axioms involving complex concept expressions. In: CEC 2020 - IEEE Congress on Evolutionary Computation, pp. 1–8. IEEE, Glasgow, United Kingdom, July 2020. https://doi.org/10.1109/CEC48606.2020.9185681
23. Nicolau, M., Agapitos, A.: Understanding grammatical evolution: grammar design, pp. 23–53, January 2018. https://doi.org/10.1007/978-3-319-78717-6_2
24. Nicolau, M., O'Neill, M., Brabazon, A.: Termination in grammatical evolution: grammar design, wrapping, and tails, pp. 1–8, June 2012. https://doi.org/10.1109/CEC.2012.6256563
25. Omran, P., Taylor, K., Rodríguez Méndez, S., Haller, A.: Learning SHACL shapes from knowledge graphs. Semant. Web **14**, 1–21 (2022). https://doi.org/10.3233/SW-223063
26. O'Neill, M., Ryan, C.: Grammatical evolution. IEEE Trans. Evol. Comput. **5**(4), 349–358 (2001)
27. Pandit, H., O'Sullivan, D., Lewis, D.: Using ontology design patterns to define SHACL shapes. In: WOP@ISWC, pp. 67–71. Monterey California, USA (2018)
28. Rabbani, K., Lissandrini, M., Hose, K.: SHACL and ShEx in the wild: a community survey on validating shapes generation and adoption. In: WWW (Companion Volume), pp. 260–263. ACM (2022)
29. Rothlauf, F., Oetzel, M.: On the locality of grammatical evolution. In: Collet, P., Tomassini, M., Ebner, M., Gustafson, S., Ekárt, A. (eds.) EuroGP 2006. LNCS, vol. 3905, pp. 320–330. Springer, Heidelberg (2006). https://doi.org/10.1007/11729976_29
30. Ryan, C., O'Neill, M., Collins, J.J.: Introduction to 20 years of grammatical evolution. In: Ryan, C., O'Neill, M., Collins, J.J. (eds.) Handbook of Grammatical Evolution, pp. 1–21. Springer, Cham (2018). https://doi.org/10.1007/978-3-319-78717-6_1

A Comprehensive Comparison
of Lexicase-Based Selection Methods
for Symbolic Regression Problems

Alina Geiger[⊠] , Dominik Sobania , and Franz Rothlauf

Johannes Gutenberg University Mainz, Mainz, Germany
{geiger,dsobania,rothlauf}@uni-mainz.de

Abstract. Lexicase selection is a parent selection method that has been
successfully used in many application domains. In recent years, several
variants of lexicase selection have been proposed and analyzed. However,
it is still unclear which lexicase variant performs best in the domain of
symbolic regression. Therefore, we compare in this work relevant lexicase
variants on a wide range of symbolic regression problems. We conduct
experiments not only over a given evaluation budget but also over a
given time as practitioners usually have limited time for solving their
problems. Consequently, this work provides users a comprehensive guide
for choosing the right selection method under different constraints in
the domain of symbolic regression. Overall, we find that down-sampled
ϵ-lexicase selection outperforms other selection methods on the studied
benchmark problems for the given evaluation budget and for the given
time. The improvements with respect to solution quality are up to 68%
using down-sampled ϵ-lexicase selection given a time budget of 24 h.

Keywords: Symbolic Regression · Genetic Programming · Lexicase
Selection

1 Introduction

Symbolic regression is an approach that aims to find a mathematical expression
that fits the points of a given data set [37]. This approach has been successfully used in a variety of applications ranging from medicine [31], and finance [3]
to even materials science [22]. Genetic Programming (GP) [26] is an evolutionary computation method commonly used to solve symbolic regression problems.
During a GP run, a randomly initialized population of individuals is gradually
improved during the evolutionary process. A recent work [13] has shown that
especially the selection process has a major impact on the solution quality.

In the last years, selection methods based on lexicase received a lot of attention due to their successful use in many application domains [1,18,30,32]. In
contrast to traditional selection methods, lexicase selection [42] considers the
error of an individual on each training case separately rather than using an

M. Giacobini et al. (Eds.): EuroGP 2024, LNCS 14631, pp. 192–208, 2024.
https://doi.org/10.1007/978-3-031-56957-9_12

aggregated fitness value. This enables lexicase selection to select specialist individuals that perform particularly well on some problem parts, which has been found to be one of the main reasons for its superiority over traditional selection methods [16,17].

Therefore, it is not surprising that several variants of lexicase selection have been proposed. In the domain of symbolic regression, ϵ-lexicase selection has shown to outperform standard lexicase selection and traditional selection methods, like tournament selection, on a wide range of regression problems [28,30]. In other domains, it has been proposed to use batches of training cases instead of single cases to evaluate the quality of individuals in the lexicase selection process [1,40]. In addition, the combination of lexicase-based selection methods with random down-sampling has been found to significantly improve the performance of GP [10,12,23]. However, which lexicase variant performs best in the domain of symbolic regression is still unclear. Answering this question would help researchers as well as practitioners to better choose an appropriate selection method for their problem at hand.

Consequently, we analyze in our work the relevant lexicase-based selection methods on a wide range of symbolic regression benchmarks by comparing the performance of each method over a given evaluation budget as well as over a given time period. As a baseline, we compare the achieved results also with tournament selection. Furthermore, we study the combination of lexicase-based selection methods with batches of training cases and with random down-sampling in the domain of symbolic regression.

We find that, given a fixed number of fitness evaluations, down-sampled ϵ-lexicase selection [12] finds significantly better solutions (with respect to the error) than all other methods; the smallest solutions are found using ϵ-plexicase selection. When studying the performance of different methods over time, performance depends on the given computation time. For example, batch-tournament selection and batch-ϵ-lexicase selection perform well if they only have around 10 min to find a solution. Moreover, variants using random down-sampling benefit from a longer search. The analysis of solutions found after 1 h and after 24 h shows that down-sampled ϵ-lexicase selection performs best overall. The improvements in terms of solution quality are up to 46% given 1 h and even up to 68% given 24 h when using down-sampled ϵ-lexicase selection. After 24 h, the differences in solution size are up to 85% with the smallest solution size found using down-sampled batch-ϵ-lexicase selection.

In summary, our main contributions are

- a comparison of the performance of relevant selection methods over a given evaluation budget and over a given time period on a wide range of symbolic regression benchmarks;
- an analysis of ϵ-lexicase selection combined with batches of training cases;
- an analysis of relevant selection methods combined with random down-sampling.

Following this introduction, Sect. 2 gives a brief overview of the related work. In Sect. 3, we describe the selection methods analyzed in this study. Section 4 presents our experimental setting and our results, followed by the conclusions in Sect. 5.

2   Related Work

We briefly discuss recent developments in the field of lexicase selection. A detailed description of lexicase-based selection methods is provided in Sect. 3.

Lexicase selection [21, 42] is a parent selection method that evaluates the quality of individuals based on their performance on each training case rather than using an aggregated fitness value. Therefore, lexicase selection selects individuals that perform extremely well on some training cases but often perform poorly on average [16, 17, 35]. In addition, prior work has found that lexicase selection is able to produce populations with a high population diversity [14, 15].

Lexicase selection has been successfully used in the domain of program synthesis [13, 18, 39, 41], symbolic regression [12, 28, 30], rule-based learning systems [1, 44], evolutionary robotics [32, 33] and even deep learning [8].

In recent years, several variants of lexicase selection emerged. First, different approaches have been proposed to relax the strict pass condition of lexicase selection to make it suitable for continuous-valued problems [30, 43], with ϵ-lexicase [30] being the most prominent one. De Melo et al. [4] introduced a method that combines tournament selection and lexicase selection to improve efficiency. Aenugu and Spector [1] proposed to use lexicase selection with batches of training cases to increase the generalizability of solutions, which has been further studied in [40]. Others have experimented with combinations of lexicase selection and novelty search to prevent premature convergence [24, 25]. In 2022, a combination of lexicase selection and weighted shuffle with partial evaluation has been proposed to improve the efficiency of lexicase selection [5, 6]. Plexicase-selection [7] is a variant that aims to improve the run time of lexicase selection by estimating the probability of individuals being selected with lexicase selection instead of actually doing the selection procedure. Variants that received a lot of attention lately combined lexicase selection with random down-sampling to explore more individuals with the same number of evaluations [2, 10, 12, 23]. Therefore, the effects of down-sampling have been further analyzed in several studies [19, 20, 38].

Prior work has compared several lexicase variants only for program synthesis [13] and not for symbolic regression. To fill this gap, we compare in this work the relevant lexicase variants on a wide range of symbolic regression problems.

3   Selection Methods

We describe tournament selection, as well as the lexicase selection variants analyzed in our study.

3.1 Tournament Selection

Tournament selection is a commonly used selection method, where individuals are compared based on an aggregated fitness measure like the mean squared error (MSE)

$$\text{MSE}(T) = \frac{1}{|T|} \sum_{t \in T} (y_t - \hat{y}_t)^2, \tag{1}$$

which is defined as the average of the squared differences between the predicted output $\hat{y}_t$ of an individual and the desired output y_t for all training cases $t \in T$.

During the selection process, tournament selection randomly chooses k individuals to participate in a tournament. The individual with the lowest error among the participants is selected as a parent [37].

3.2 ϵ-Lexicase Selection

Evaluating individuals based on a single fitness value leads to a loss of information regarding the structure of the data [27]. Therefore, lexicase selection has been proposed that considers the performance of individuals on each training case separately [42]. For continuous-valued problems, ϵ-lexicase selection [30] has been found to perform well due to the introduction of an ϵ-threshold that relaxes the strict pass condition of lexicase selection. The ϵ-lexicase selection algorithm is shown in Algorithm 1 for a training set T and a population P according to La Cava et al. [30].

Algorithm 1 ϵ-Lexicase Selection

1: $C :=$ all individuals $i \in P$
2: $T' :=$ randomly shuffled T
3: **while** $|T'| > 0$ and $|C| > 1$ **do**
4:     $t :=$ first case of T'
5:     $\lambda_t := e_t^* + \epsilon_t$
6:     $C :=$ all candidates with $e_t <= \lambda_t$
7:     **remove** t from T'
8: **end while**
9: **return** random choice from C

For each selection event, all individuals $i \in P$ are considered as candidates $c \in C$ for selection and all training cases $t \in T$ are randomly shuffled. Next, ϵ-lexicase selection iterates through all training cases $t \in T'$, keeping in each iteration only the candidates $c \in C$ with an error e_t lower than or equal to the threshold λ_t, where λ_t is defined as the sum of the lowest error on the current case e_t^* and ϵ_t. As suggested by La Cava et al. [28], ϵ_t can be dynamically calculated for a training case t using the median absolute deviation [36]

$$\epsilon_t = \text{median}(|\mathbf{e}_t - \text{median}(\mathbf{e}_t)|), \tag{2}$$

where $\mathbf{e}_t$ is a vector of all errors across the candidates $c \in C$ on case t. The iterative process ends if there is only one candidate left or if there are no more training cases. In case there is more than one candidate left, one is randomly selected as a parent from C.

3.3   Batch-ϵ-Lexicase Selection

When using lexicase, the selection depends on the ordering of training cases [1]. Therefore, Aenugu and Spector [1] have proposed to use batches of training cases instead of single training cases to evaluate individuals in the lexicase selection process. It has been found that the use of batches can increase the generalizability of the generated solutions [1, 40]. In this work, we propose to combine ϵ-lexicase selection with batches of training cases for solving symbolic regression problems. In each generation, the training cases are randomly combined to batches with a batch size b. The selection is performed according to Algorithm 1 with the modification that individuals are evaluated based on their performance (e.g., in terms of the MSE) on each batch.

3.4   Batch-Tournament Selection

One drawback of lexicase-based methods is that individuals are compared on a usually large number of training cases which is computationally expensive [4]. Therefore, batch-tournament selection [4] has been proposed to combine the advantages of lexicase selection and tournament selection. Similar to batch-ϵ-lexicase selection, the training cases are randomly combined to batches[1] of size b. For each selection, batch-tournament selection chooses k individuals to participate in a tournament. The participant with the lowest error on the current batch is selected as a parent. For each selection, another batch is used to compare the participants in the tournament. If there are more selections than batches, the batches are used repeatedly.

3.5 ϵ-Plexicase Selection

A recent approach to reduce the run time of lexicase selection has been proposed by Ding et al. [7] in 2023. They suggest to approximate the probability of individuals being selected with ϵ-lexicase selection and to select parents based on the probability distribution instead of doing the actual selection procedure described in Algorithm 1. Their method, ϵ-plexicase selection, requires a parameter α that can be used to manipulate the probability distribution. Ding et al. [7] have found that ϵ-plexicase selection is significantly more efficient than ϵ-lexicase selection. For a more detailed description of ϵ-plexicase selection we refer to [7].

---

[1] We refer to the variant called BTSS from De Melo et al. [4].

3.6   Combining Lexicase Variants with Random Down-Sampling

Usually, the search process is limited by a computational budget as computational effort increases with the size of the training set [23]. Therefore, Hernandez et al. [23] proposed to use only a random subset of the training cases in each generation for the evaluation of individuals to be able to search for more generations with the same evaluation budget. In the domain of symbolic regression, the combination of random down-sampling with ϵ-lexicase selection has been found to significantly outperform ϵ-lexicase selection on a range of benchmark problems [12]. Therefore, we combine in our study all selection methods described above with random down-sampling to further explore its benefits.

4   Experiments

The aim of this work is to provide users with a guideline for choosing an appropriate selection method for symbolic regression problems. Therefore, we compare the selection methods described in Sect. 3, as well as their down-sampled variants on a wide range of symbolic regression benchmarks by analyzing their performance over a given number of evaluations, as well as over a given time period.

In this section, we present our experimental setup, e.g., the benchmark problems used and the parameter choice of our GP approach. In addition, we present and discuss the results.

4.1   Experimental Setting

We evaluate the different selection methods on 20 symbolic regression benchmark problems[2] that we sampled without replacement from SR bench [29]. We only consider problems with less than or equal to 1,000 observations to ensure that all experiments for the given evaluation budget finish within 24 h. The number of observations range from 100 to 1,000 and the number of input features range from 2 to 50 for the chosen problems.

We use the DEAP framework [11] (version 1.4.1) for our implementation. The parameter setting of our GP approach is shown in Table 1, following the approach of Geiger et al. [12].

The primitive set consists of the input features x, an ephemeral random constant (ERC) that can take the values -1, 0, or 1, and the functions for addition, subtraction, multiplication, analytic quotient (AQ) [34], sine, cosine, and a function that returns the negative of the given value. We set the population size to 500 and initialize the population using ramped half-and-half with tree

---

[2] problems: 589_fri_c2_1000_25, 606_fri_c2_1000_10, 623_fri_c4_1000_10, 1030_ERA, 607_fri_c4_1000_50, 581_fri_c3_500_25, 617_fri_c3_500_5, 654_ fri_c0_500_10, 641_fri_c1_500_10, 1027_ESL, 519_vinnie, 647_fri_c1_250_ 10, 615_fri_c4_250_10, 230_machine_cpu, 207_autoPrice, 665_sleuth_case2002, 523_analcatdata_neavote, 621_fri_c0_100_10, 624_fri_c0_100_5, 591_fri_c1_ 100_10.

Table 1. GP Parameter Setting

Parameter	Value
Population size	500
Primitive set	$\{\mathbf{x}, \text{ERC}, +, -, *, \text{AQ}, \sin, \cos, \text{neg}\}$
ERC values	$\{-1, 0, 1\}$
Initialization method	Ramped half-and-half
Maximum tree depth	17
Crossover probability	80%
Mutation probability	5%
Runs	30

depths ranging from 0 to 4. During a run, the maximum tree depth is restricted to 17 [26]. Crossover is applied with a probability of 80% and mutation with a probability of 5%.

The parameters of the selection methods are carefully chosen taking into account the results of previous studies. For tournament selection, we set a tournament size $k = 5$ which is within the range of commonly used tournament sizes [9]. For batch-tournament selection, we set a tournament size $k = 64$, which has been found to perform well [4]. For batch-tournament selection and batch-ϵ-lexicase selection, we study batch sizes b of $0.05|T|$, $0.075|T|$ and $0.1|T|$, where $|T|$ is the number of training cases [1]. This means for a batch size $b = 0.05|T|$, a single batch consists of 5% of the available training cases. For ϵ-plexicase selection, we set $\alpha = 1.0$ [7]. For the down-sampled variants, we set a down-sampling rate $d = 0.1$ as suggested by [12], meaning that only 10% of the training cases are used per generation to evaluate individuals.

Before each run, we randomly split the data set for each problem into 70% training cases, 15% validation cases, and 15% test cases. The training cases are used to evaluate the quality of individuals during the search. The definition of the fitness function depends on the chosen selection method. For tournament selection, we use the MSE as individuals are compared based on an aggregated fitness measure. In contrast, ϵ-lexicase selection and ϵ-plexicase selection consider the performance of individuals on each training case separately. Therefore, we use a fitness function that returns for an individual its squared errors on each training case. Batch-ϵ-lexicase selection and batch-tournament selection evaluate individuals on batches of training cases rather than single training cases. For both methods, we define a fitness function that returns an individual's MSE on each batch.

Each GP run returns the individual with the lowest MSE on the validation cases (that have never been used for training). Therefore, we avoid returning a solution that is overfitted to the training cases. The quality of the returned solution is then evaluated in terms of its MSE on the unseen test cases.

We analyze the performance of each method over a given evaluation budget, as this is usually used to compare selection methods [12, 19]. The evaluation budget is defined as the population size multiplied by the number of training cases and the number of generations. For the variants without down-sampling we give each method 100 generations to search for a solution. The down-sampling variants use the same evaluation budget which means they can search for 1,000 generations instead of 100, as only 10% of the training cases are used in each generation.

In addition, we observe the performance of each selection method over 86, 400 seconds (24 h). In each generation, we track the lowest MSE on the validation cases measured for an individual in the population, the average size of the trees (in terms of number of nodes), and the time. Due to space limitations, we compare the results across all problems in an aggregated form. Therefore, we normalized the MSE measured on the test and validation cases for each problem using a min-max normalization. We refer to the normalized MSE as MSE′.

For each selection method, parameter setting, and problem we perform 30 runs. The experiments were performed on a high performance computing cluster using Intel 2630v4 2,20GHz CPUs. All processes were single threaded. We measured the time in terms of wall clock time.

4.2   Results

We compare the performance of ϵ-lexicase selection (denoted as ϵ-lex), ϵ-plexicase selection (ϵ-plex), tournament selection (tourn), batch-ϵ-lexicase selection (batch-ϵ-lex), batch-tournament selection (batch-tourn), and their down-sampling variants (denoted by the prefix down). We only show the results for the best batch size found for each problem, meaning the batch size that achieved on average the lowest MSE on the validation cases.

First, we compare the methods given a fixed evaluation budget as this is usually used to compare selection methods in the literature [12, 19]. Figure 1 shows the mean normalized MSE on the validation cases (for the best-performing individual in each generation) across all problems over the normalized evaluation budget for each selection method. A normalized evaluation budget of 1.0 refers to 100 generations for selection methods without down-sampling and 1, 000 generations for selection methods with down-sampling. We can observe that down-ϵ-lex performs best overall. At the beginning, down-batch-tourn is the second-best performing method. However, this changes over time with ϵ-lex being the second-best method in the end. Tourn and ϵ-plex achieve the highest errors in terms of the MSE.

We analyze the solutions found within the given evaluation budget in detail in Table 2. We report for each selection method the average rank $\bar{r}$ and the average normalized MSE on the unseen test cases $\overline{MSE'}$, as well as the average solution size $\bar{s}$, the number of generations $\bar{g}$, and the average duration in seconds $\bar{\theta}$ across 20 problems. Note that the reported ranks and errors might differ from the ones observed in Fig. 1 as they refer to the final performance on the unseen test cases (not the performance on the validation cases). We test whether the solution

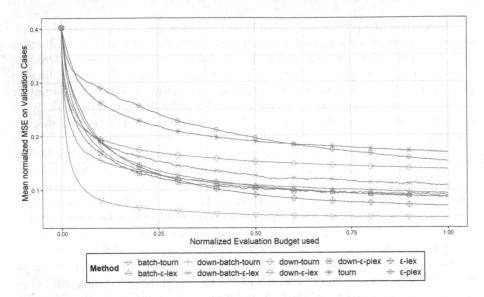

Fig. 1. Mean normalized MSE on the validation cases across all 20 problems over a given evaluation budget for different selection methods. For better readability, markers are added to each line.

quality of the selection methods differs significantly by performing a pairwise Wilcoxon rank-sum test with a Holm correction. Significant improvements ($p <$ 0.05) with respect to other selection methods are indicated by the methods label $a - j$.

Down-ϵ-lex achieves the best average rank across all problems ($\bar{r} = 3.30$) and significantly outperforms all other methods in terms of the average MSE on the test cases. Analyzing the performance on each individual problem shows that down-ϵ-lex outperforms all other selection methods on 12 out of 20 problems. The next best performing methods are ϵ-lex ($\bar{r} = 4.40$), batch-ϵ-lex ($\bar{r} = 4.80$), and down-ϵ-plex ($\bar{r} = 4.80$). The size of the final solution varies from $\bar{s} = 72$ using ϵ-plex to $\bar{s} = 265$ using down-tourn. Interestingly, ϵ-plex only takes 530 seconds to perform the search for the given evaluation budget. Batch-tourn and tourn are also relatively fast. Down-ϵ-lex takes 12, 480 seconds which is more than 23 times the duration of ϵ-plex.

Figure 2 plots the average solution size over time for each selection method. We can observe an increase in solution size over time for all methods, with the largest increase using down-tourn. As evaluating larger solutions is more expensive and evaluating generations usually gets more expensive as solutions grow over time, it is important to also take into account the time and not only the number of evaluations. Moreover, we have seen in Table 2 that the time complexity of the selection methods varies greatly, which means that some methods could perform more generations in the same amount of time than others.

Thus, we also compare the selection methods over a given time period. Figure 3 shows the mean normalized MSE on the validation cases across all prob-

Table 2. Average rank $\bar{r}$ of each method, average normalized MSE on the test cases $\overline{\text{MSE}'}$, average solution size $\bar{s}$, number of generations $\bar{g}$, and average duration in seconds $\bar{\theta}$ across all 20 problems **for a given evaluation budget**. Significant improvements in terms of solution quality are denoted by the method labels $a - j$. Best results are highlighted in bold font.

Method	$\bar{r}$	$\overline{\text{MSE}'}$	$\bar{s}$	$\bar{g}$	$\bar{\theta}$
[a]batch-tourn	5.85	[hj]0.0552	102	100	656
[b]batch-ε-lex	4.80	[hj]0.0497	135	100	1,192
[c]down-batch-tourn	5.00	[hj]0.0508	123	1,000	3,633
[d]down-batch-ε-lex	5.30	[hj]0.0526	95	1,000	3,680
[e]down-tourn	6.10	0.0666	265	1,000	6,432
[f]down-ε-lex	**3.30**	[abcdeghij]**0.0288**	135	1,000	12,480
[g]down-ε-plex	4.80	[hj]0.0509	102	1,000	2,867
[h]tourn	7.25	0.0739	133	100	785
[i]ε-lex	4.40	[ehj]0.0461	99	100	8,320
[j]ε-plex	8.20	0.0798	**72**	100	530

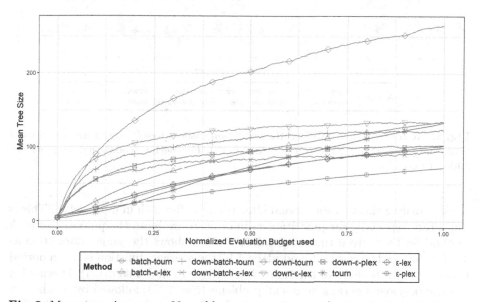

Fig. 2. Mean tree size across 20 problems over a given evaluation budget. For better readability, markers are added to each line.

lems over 24 h for each selection method.[3] We zoom in over the first hour, as we observe most changes in performance in this time period. Batch-tourn achieves

---

[3] We track in each generation the MSE of the best-performing individual (according to the performance on the validation cases) in the current population. The normalized MSE is averaged over all problems every 10 min.

the lowest MSE on the validation cases after around 10 min, followed by batch-ϵ-lex and down-ϵ-lex, while ϵ-lex performs worst. However, down-ϵ-lex outperforms batch-tourn after around 30 min. At this point in time, the worst performance is observed with tourn and down-tourn. Down-ϵ-lex remains the best performing method in the remaining time period. ϵ-lex benefits from a longer search as it changes from being the worst selection method to being the second-best over 24 h.

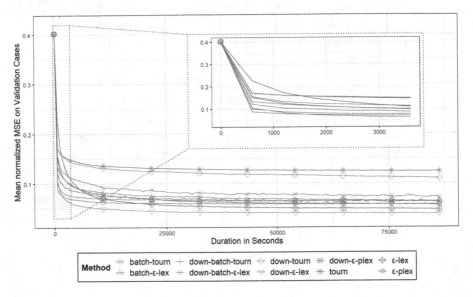

Fig. 3. Mean normalized MSE on the validation cases across 20 problems over a fixed time budget of 24 h (zoom over the first hour). For better readability, markers are added to each line.

We analyze the solutions found after 1 h and after 24 h in more detail. This is important especially for practitioners in order to evaluate the potential of each method for their given time constraints. Table 3 shows the same information as Table 2 but for the results found **after 1 h** (the average duration is not reported because all methods have the same time budget to find a solution). Down-ϵ-lex has the best average rank across all problems ($\bar{r} = 3.45$), followed by batch-ϵ-lex ($\bar{r} = 3.60$) and batch-tourn ($\bar{r} = 4.05$). In detail, down-ϵ-lex outperforms all other selection methods on 9 out of 20 problems. Down-ϵ-lex performs significantly better than down-tourn, tourn, ϵ-lex, and ϵ-plex with differences up to 46% in terms of the MSE on the test cases. Batch-tourn and batch-ϵ-lex significantly outperform down-tourn, tourn, and ϵ-lex. There are also major differences in the average solution size and the average number of generations performed within the given time period. For example, the smallest average solution size is found using ϵ-lex ($\bar{s} = 79$), and the largest average solution size is found using tourn ($\bar{s} = 287$). The differences in the average number of generations are up to 96%, with the smallest number observed using ϵ-lex ($\bar{g} = 77$), and the largest using down-ϵ-plex ($\bar{g} = 2,438$).

Table 3. Average rank $\bar{r}$ of each method, average normalized MSE on the test cases $\overline{\text{MSE}'}$, average solution size $\bar{s}$, and average number of generations $\bar{g}$ across all 20 problems **after 1 h**. Significant improvements with in terms of solution quality are denoted by the method labels $a - j$. Best results are highlighted in bold font.

Method	$\bar{r}$	$\overline{\text{MSE}'}$	$\bar{s}$	$\bar{g}$
[a]batch-tourn	4.05	[ehi]0.0435	177	559
[b]batch-ϵ-lex	3.60	[ehi]0.0420	203	334
[c]down-batch-tourn	4.85	[eh]0.0478	124	2,052
[d]down-batch-ϵ-lex	5.30	0.0544	99	1,328
[e]down-tourn	7.40	0.0686	243	1,086
[f]down-ϵ-lex	**3.45**	[ehij]**0.0368**	115	449
[g]down-ϵ-plex	5.35	[e]0.0502	110	2,438
[h]tourn	6.75	0.0675	287	422
[i]ϵ-lex	7.15	0.0603	**79**	77
[j]ϵ-plex	7.10	0.0564	133	572

Table 4 shows details for the solutions found after **after 24 h**. Down-ϵ-lex achieves the best average rank ($\bar{r} = 3.10$) and significantly outperforms 8 out of 9 selection methods with improvements up to 68% in comparison to the baseline (tourn). We observed that down-ϵ-lex outperforms all other methods on 12 out of 20 problems. Apart from that, there are several changes in the method ranks compared to Table 3. Down-ϵ-plex is now the second-best performing method ($\bar{r} = 4.55$), followed by down-batch-tourn ($\bar{r} = 4.60$) and down-batch-ϵ-lex ($\bar{r} = 4.75$). In comparison to the results in Table 3, all down-sampled variants achieve a lower rank than before. This indicates that down-sampling benefits from having more time. We want to note, that some results differ from the observations made in Fig. 3, where, e.g., ϵ-lex has been the second-best performing method in terms of the MSE on the validation cases, indicating that the ability to generalize to unseen data differs between methods.

All methods find solutions with a larger average size compared to the solutions found after 1 h (see Table 3). The average solution size is smallest using down-batch-ϵ-lex ($\bar{s} = 120$), followed by down-ϵ-plex ($\bar{s} = 136$), and down-batch-tourn ($\bar{s} = 153$). The largest average solution size is observed using tourn ($\bar{s} = 850$). Thus, the differences in solution size are up to 85%. Generally, we note that the versions using down-sampling generate smaller solutions than their counterpart without down-sampling. Moreover, the down-sampled versions are able to pass through more generations, because only a fraction of the training cases is used to evaluate individuals in each generation.

Table 4. Average rank $\bar{r}$ of each method, average normalized MSE on the test cases $\overline{MSE}'$, average solution size $\bar{s}$, and average number of generations $\bar{g}$ across all 20 problems **after 24 h**. Significant improvements with in terms of solution quality are denoted by the method labels $a - j$. Best results are highlighted in bold font.

Method	$\bar{r}$	$\overline{MSE}'$	$\bar{s}$	$\bar{g}$
[a]batch-tourn	6.20	[eh]0.0322	290	8,221
[b]batch-ε-lex	6.25	[eh]0.0329	443	4,615
[c]down-batch-tourn	4.60	[eh]0.0277	153	38,990
[d]down-batch-ε-lex	4.75	[eh]0.0262	**120**	27,199
[e]down-tourn	7.00	0.0550	460	14,495
[f]down-ε-lex	**3.10**	[abceghij]**0.0181**	158	9,401
[g]down-ε-plex	4.55	[eh]0.0295	136	52,204
[h]tourn	7.50	0.0579	850	3,506
[i]ε-lex	5.30	[eh]0.0286	213	1,517
[j]ε-plex	5.75	[eh]0.0300	207	8,570

To sum up: given a fixed evaluation budget, down-ε-lex significantly outperforms all other methods with respect to the MSE of the found solutions and ε-plex generates on average the smallest solutions. Additionally, performance depends on the time budget. This is relevant for practitioners as they often have limited time to solve their regression problems. We found that batch-tourn and batch-ε-lex perform best after around 10 min in terms of the MSE on the validation cases. This changes after around 30 min with down-ε-lex being the best-performing method. The analysis of solutions found after 24 h shows that down-ε-lex significantly outperforms 8 out of 9 studied methods in terms of solution quality on the unseen test cases. After 24 h, smallest solutions are found with down-batch-ε-lex.

5   Conclusions

We compared the relevant lexicase variants on a wide range of symbolic regression problems taken from SR Bench to provide users with a comprehensive guide for choosing the appropriate selection method for their problem at hand. We analyzed each selection method over a given evaluation budget (number of evaluations), as well as over a given time period. Furthermore, we studied the influence of each selection method on the solution size.

We found that down-sampled ε-lexicase selection significantly outperforms all other selection methods with respect to the MSE and that the smallest solution size is found using ε-plexicase selection for a given evaluation budget. Additionally, we observed that the relative performance of each method depends on the given time period. For example, we observed that batch-tournament selection and batch-ϵ-lexicase selection perform well (in terms of the MSE on the validation cases) in comparison to other selection methods if the given time period is very short (around 10 min). However, the detailed analysis of the solutions found after 1 h or after 24 h reveals that down-sampled ϵ-lexicase selection performs best under these conditions. The improvements with respect to the solution quality on the unseen test cases are up to 46% using down-sampled ϵ-lexicase selection for a given time period of 1 h. Given 24 h, the differences are even up to 68%. Moreover, we find that lexicase variants using random down-sampling benefit from a longer search. In terms of solution size, we observe differences up to 85% with the smallest solutions found using down-sampled batch-ϵ-lexicase selection for a given time period of 24 h. All in all, we recommend down-sampled ϵ-lexicase selection as a parent selection method for solving symbolic regression problems.

In future work, we will analyze if the performance of the selection methods depends on certain characteristics of the data set, e.g., the number of observations or the number of features. Therefore, we will show the performance on each benchmark problem separately. Additionally, the influence of the selection methods on population dynamics such as population diversity or specialist selection could be studied.

References

1. Aenugu, S., Spector, L.: Lexicase selection in learning classifier systems. In: Proceedings of the Genetic and Evolutionary Computation Conference, pp. 356–364. ACM (2019)
2. Boldi, R., et al.: Informed down-sampled lexicase selection: Identifying productive training cases for efficient problem solving. arXiv preprint arXiv:2301.01488v1 (2023)
3. Chen, S.H.: Genetic Algorithms and Genetic Programming in Computational Finance. Springer, New York (2012). https://doi.org/10.1007/978-1-4615-0835-9
4. De Melo, V.V., Vargas, D.V., Banzhaf, W.: Batch tournament selection for genetic programming: the quality of lexicase, the speed of tournament. In: Proceedings of the Genetic and Evolutionary Computation Conference, pp. 994–1002. GECCO '19, ACM (2019)
5. Ding, L., Boldi, R., Helmuth, T., Spector, L.: Going faster and hence further with lexicase selection. In: Proceedings of the Genetic and Evolutionary Computation Conference Companion, pp. 538–541. ACM (2022)
6. Ding, L., Boldi, R., Helmuth, T., Spector, L.: Lexicase selection at scale. In: Proceedings of the Genetic and Evolutionary Computation Conference Companion, pp. 2054–2062. ACM (2022)
7. Ding, L., Pantridge, E., Spector, L.: Probabilistic lexicase selection. In: Proceedings of the Genetic and Evolutionary Computation Conference, pp. 1073–1081. GECCO '23, ACM (2023)
8. Ding, L., Spector, L.: Optimizing neural networks with gradient lexicase selection. In: International Conference on Learning Representations (2021)

9. Fang, Y., Li, J.: A review of tournament selection in genetic programming. In: Cai, Z., Hu, C., Kang, Z., Liu, Y. (eds.) Advances in Computation and Intelligence. ISICA 2010. LNCS, vol. 6382, pp. 181–192. Springer, Berlin, Heidelberg (2010). https://doi.org/10.1007/978-3-642-16493-4_19

10. Ferguson, A.J., Hernandez, J.G., Junghans, D., Lalejini, A., Dolson, E., Ofria, C.: Characterizing the effects of random subsampling on lexicase selection. In: Banzhaf, W., Goodman, E., Sheneman, L., Trujillo, L., Worzel, B. (eds.) Genetic Programming Theory and Practice XVII. Genetic and Evolutionary Computation, LNCS, pp. 1–23. Springer, Cham (2020). https://doi.org/10.1007/978-3-030-39958-0_1

11. Fortin, F.A., de Rainville, F.M., Gardner, M.A., Parizeau, M., Gagné, C.: DEAP: evolutionary algorithms made easy. J. Mach. Learn. Res. **13**(1), 2171–2175 (2012)

12. Geiger, A., Sobania, D., Rothlauf, F.: Down-sampled epsilon-lexicase selection for real-world symbolic regression problems. In: Proceedings of the Genetic and Evolutionary Computation Conference, pp. 1109–1117. GECCO '23, ACM (2023)

13. Helmuth, T., Abdelhady, A.: Benchmarking parent selection for program synthesis by genetic programming. In: Proceedings of the 2020 Genetic and Evolutionary Computation Conference Companion, pp. 237–238. GECCO '20, ACM (2020)

14. Helmuth, T., McPhee, N.F., Spector, L.: Effects of lexicase and tournament selection on diversity recovery and maintenance. In: Proceedings of the 2016 on Genetic and Evolutionary Computation Conference Companion, pp. 983–990. GECCO '16 Companion, ACM (2016)

15. Helmuth, T., McPhee, N.F., Spector, L.: Lexicase selection for program synthesis: a diversity analysis. In: Riolo, R., Worzel, W., Kotanchek, M., Kordon, A. (eds.) Genetic Programming Theory and Practice XIII, pp. 151–167. LNCS, Genetic and Evolutionary Computation. Springer, Cham (2016). https://doi.org/10.1007/978-3-319-34223-8_9

16. Helmuth, T., Pantridge, E., Spector, L.: Lexicase selection of specialists. In: Proceedings of the Genetic and Evolutionary Computation Conference, pp. 1030–1038. GECCO '19, ACM (2019)

17. Helmuth, T., Pantridge, E., Spector, L.: On the importance of specialists for lexicase selection. Genet. Program. Evolvable Mach. **21**(3), 349–373 (2020)

18. Helmuth, T., Spector, L.: General program synthesis benchmark suite. In: Proceedings of the 2015 Annual Conference on Genetic and Evolutionary Computation, pp. 1039–1046. GECCO '15, ACM (2015)

19. Helmuth, T., Spector, L.: Explaining and exploiting the advantages of down-sampled lexicase selection. In: ALIFE 2020: The 2020 Conference on Artificial Life, pp. 341–349. MIT Press (2020)

20. Helmuth, T., Spector, L.: Problem-solving benefits of down-sampled lexicase selection. Artif. Life **27**(3–4), 183–203 (2021)

21. Helmuth, T., Spector, L., Matheson, J.: Solving uncompromising problems with lexicase selection. IEEE Trans. Evol. Comput. **19**(5), 630–643 (2014)

22. Hernandez, A., Balasubramanian, A., Yuan, F., Mason, S.A., Mueller, T.: Fast, accurate, and transferable many-body interatomic potentials by symbolic regression. NPJ Comput. Mater. **5**(1), 112 (2019)

23. Hernandez, J.G., Lalejini, A., Dolson, E., Ofria, C.: Random subsampling improves performance in lexicase selection. In: Proceedings of the Genetic and Evolutionary Computation Conference Companion, pp. 2028–2031. GECCO '19, ACM (2019)

24. Jundt, L., Helmuth, T.: Comparing and combining lexicase selection and novelty search. In: Proceedings of the Genetic and Evolutionary Computation Conference, pp. 1047–1055. ACM (2019)

25. Kelly, J., Hemberg, E., O'Reilly, U.M.: Improving genetic programming with novel exploration - exploitation control. In: Sekanina, L., Hu, T., Lourenco, N., Richter, H., Garcia-Sanchez, P. (eds.) Genetic Programming. EuroGP 2019. LNCS, vol. 11451, pp. 64–80. Springer, Cham (2019). https://doi.org/10.1007/978-3-030-16670-0_5

26. Koza, J.R.: On the Programming of Computers by Means of Natural Selection, A Bradford Book, vol. 1. MIT Press, Cambridge (1992)

27. Krawiec, K., Liskowski, P.: Automatic derivation of search objectives for test-based genetic programming. In: Machado, P., et al. (eds.) Genetic Programming. EuroGP 2015. LNCS, vol. 9025, pp. 53–65. Springer, Cham (2015). https://doi.org/10.1007/978-3-319-16501-1_5

28. La Cava, W., Helmuth, T., Spector, L., Moore, J.H.: A probabilistic and multi-objective analysis of lexicase selection and epsilon-lexicase selection. Evol. Comput. **27**(3), 377–402 (2019)

29. La Cava, W., et al.: Contemporary symbolic regression methods and their relative performance. In: Thirty-fifth Conference on Neural Information Processing Systems Datasets and Benchmarks Track (2021)

30. La Cava, W., Spector, L., Danai, K.: Epsilon-lexicase selection for regression. In: Proceedings of the Genetic and Evolutionary Computation Conference 2016, pp. 741–748. GECCO '16, ACM (2016)

31. La Cava, W.G., et al.: A flexible symbolic regression method for constructing interpretable clinical prediction models. NPJ Digit. Med. **6**(1), 107 (2023)

32. Moore, J.M., Stanton, A.: Lexicase selection outperforms previous strategies for incremental evolution of virtual creature controllers. In: ECAL 2017, The Fourteenth European Conference on Artificial Life, pp. 290–297 (2017)

33. Moore, J.M., Stanton, A.: Tiebreaks and diversity: isolating effects in lexicase selection. In: The 2018 Conference on Artificial Life, pp. 590–597. MIT Press, Cambridge, MA (2018)

34. Ni, J., Drieberg, R.H., Rockett, P.I.: The use of an analytic quotient operator in genetic programming. IEEE Trans. Evol. Comput. **17**(1), 146–152 (2013)

35. Pantridge, E., Helmuth, T., McPhee, N.F., Spector, L.: Specialization and elitism in lexicase and tournament selection. In: Proceedings of the Genetic and Evolutionary Computation Conference Companion, pp. 1914–1917. GECCO '18, ACM (2018)

36. Pham-Gia, T., Hung, T.L.: The mean and median absolute deviations. Math. Comput. **34**(7–8), 921–936 (2001)

37. Poli, R., Langdon, W.B., McPhee, N.F.: A Field Guide to Genetic Programming. Lulu Press, Morrisville (2008)

38. Schweim, D., Sobania, D., Rothlauf, F.: Effects of the training set size: a comparison of standard and down-sampled lexicase selection in program synthesis. In: 2022 IEEE Congress on Evolutionary Computation (CEC), pp. 1–8. IEEE (2022)

39. Sobania, D., Rothlauf, F.: A generalizability measure for program synthesis with genetic programming. In: Proceedings of the Genetic and Evolutionary Computation Conference, pp. 822–829. GECCO '21, ACM (2021)

40. Sobania, D., Rothlauf, F.: Program synthesis with genetic programming: the influence of batch sizes. In: Medvet, E., Pappa, G., Xue, B. (eds.) Genetic Programming. EuroGP 2022. LNCS, vol. 13223, pp. 118–129. Springer, Cham (2022). https://doi.org/10.1007/978-3-031-02056-8_8

41. Sobania, D., Schweim, D., Rothlauf, F.: A comprehensive survey on program synthesis with evolutionary algorithms. IEEE Trans. Evol. Comput. **27**(1), 82–97 (2023)

42. Spector, L.: Assessment of problem modality by differential performance of lexicase selection in genetic programming: a preliminary report. In: Proceedings of the 14th Annual Conference Companion on Genetic and Evolutionary Computation, pp. 401–408. GECCO '12, ACM (2012)

43. Spector, L., La Cava, W., Shanabrook S, Helmuth, T., Pantridge, E.: Relaxations of lexicase parent selection. In: Banzhaf, W., Olson, R., Tozier, W., Riolo, R. (eds.) Genetic Programming Theory and Practice XV, pp. 105–120. LNCS, Genetic and Evolutionary Computation. Springer, Cham (2018). https://doi.org/10.1007/978-3-319-90512-9_7

44. Wagner, A.R.M., Stein, A.: Adopting lexicase selection for michigan-style learning classifier systems with continuous-valued inputs. In: Proceedings of the Genetic and Evolutionary Computation Conference Companion, pp. 171–172. ACM (2021)

Genetic Improvement of Last Level Cache

William B. Langdon$^{(\boxtimes)}$ and David Clark

CREST, Department of Computer Science, UCL,
Gower Street, London WC1E 6BT, UK
W.Langdon@cs.ucl.ac.uk, david.clark@ucl.ac.uk
http://www.cs.ucl.ac.uk/staff/W.Langdon,http://crest.cs.ucl.ac.uk/ ,
http://www.cs.ucl.ac.uk/staff/D.Clark

Abstract. With increasing reliance on multi-core parallel computing performance is evermore dominated by interprocessor data communication typically provided by last level cache (LLC) shared between CPUs. In an 8 core 3.6 GHz desktop using multiple local searches, the Magpie parameter tuning genetic improvement (GI) system was able to reduce L3 cache access (load + stores) four fold on an existing open source 7000 line C PARSEC parallel computing VIPS image benchmark.

Keywords: hill climbing · SBSE, Software Engineering · automatic code optimisation · srcml · XML · parameter tuning · reduced search space · Linux perf

1   Introduction

The computing industry grew up in the presence of Moore's Law [44] ensuring in the early days software producers could always access more powerful computers by the time their programs were ready for release. The days of clock speeds doubling every two years are long gone, however silicon chip manufactures are using the still increasing number of transistors to pack more processing cores and larger cache memory onto their devices. This will continue into the foreseeable future with more cores becoming available and so increasing importance of communications between CPUs. Excluding specialised hardware, such as FPGAs and GPUs, many current parallel applications communicate between CPUs via shared memory. In multi-core silicon chips this can be highly effective. However modern CPUs run far faster than main memory and a complex hierarchy of cache memory is needed to try to keep data close to the individual computing engines. In many parallel multi-threaded applications the last level cache (LLC) is shared between cores and provides the main communication between threads running on different CPUs within the same chip. In most cases control of the cache hierarchy remains proprietary. Although cache memory will increase in size, it will remain both the main bottleneck limiting performance for many applications and outside programmer control. We show genetic improvement can in principle be used to automatically tune open source software to minimise use

© The Author(s), under exclusive license to Springer Nature Switzerland AG 2024
M. Giacobini et al. (Eds.): EuroGP 2024, LNCS 14631, pp. 209–226, 2024.
https://doi.org/10.1007/978-3-031-56957-9_13

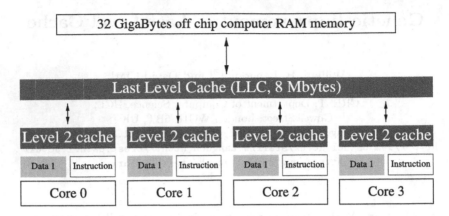

Fig. 1. Schematic of three level on-chip cache hierarchy. The last level (here level 3) cache is by far the largest on chip cache. (In our desktop each data and instruction L1 cache is 32 Kbytes, L2 each 256 Kbytes and the LLC (L3) is 8 Mbytes.) As well as interfacing with off chip memory, LLC also provides communication between the compute cores (just 4 cores are shown).

of the shared LLC cache, obtaining a 4.0× reduction, without deep access to the operating system or inner workings of the silicon chip.

PARSEC (Princeton Application Repository for Shared-Memory Computers) is a benchmark suite of parallel computing programs, which focuses on emerging workloads [5, page 73]. It includes VIPS [43], which is an image processing library written in C. We selected the VIPS thumbnail image processing benchmark as it is multi-threaded and can be easily scaled to cover the critical size of modern chip caches. Indeed we use the Linux perf tool to measure its cache use during its multi-threaded generation of a small "thumbnail" image from an image exceeding the cache size (see Fig. 2 https://github.com/wblangdon/vips). To avoid potentially complicated trade-offs between cache, image size and image quality (available in much more complicated image formats such as JPEG), for both images we use non-compressed P6 raster scan full colour images (see Sect. 7.3) and insist the mutated code produces identical output.

Fig. 2. 128 × 96 thumbnail image generated by VIPS.

In Sect. 3 we describe our use of the Magpie [8] genetic improvement system to simultaneously tune application specific parameters, compiler and linker options, and the VIPS C source code. The VIPS benchmark is detailed in Sect. 4. Whilst Sect. 5 describes how we use the Linux perf API to measure last level cache LLC usage and measures to combat noise. Section 6 shows in many cases Magpie is able to reduce cache usage by on average 75% ($\pm 4\%$). In the discussion (Sect. 7) we note that, amongst other changes, most successful mutations involved VIPS application parameters and (in Sect. 7.2) run and report additional experiments just tuning them. Section 7.3 summarises the VIPS thumbnail code and proposes an explanation for why Magpie's patches work. We conclude (Sect. 8) that despite noise, Magpie can find a single parameter change which reduces LLC cache use four fold. But first we give the background.

2   Background

Until recently genetic improvement (GI) [25,47] has applied genetic programming (GP) [3,21,48], to existing human written software, however in principle any optimisation technique, such as search-based software engineering [14], Grammatical Evolution [15–45], Novelty Search [13], Fuzzy Systems [57], or AI [10–56], can be used. Recently Magpie [8] has shown the power of local search in GI [7,9,27]. Already genetic improvement has been applied to automatic porting [25], transplanting code [41,42] code optimisation [6,32], including JavaScript [12] and Clang LLVM IR intermediate code [26,52], hardware design [11], automatic software testing [45] and cryptographic code [23]. Genetic improvement has been demonstrated on GPU applications [30–38] including BarraCUDA [20], the first GI code to be accepted into actual use [24]. At EuroGP'19 [34] we showed GI could also speed up parallel CPU code. The resulting GIed RNAfold [39] was accepted into production and like the GI version of BarraCUDA has been downloaded many thousands of times (for example [2]).

Previously we [36] showed GP optimising L1 cache but L1 is much more tightly bound to the CPU running the application (see Fig. 1) and in half the cases we were unable to find an improvement. Jimenez et al. [17] use a genetic algorithm (GA) to improve the LLC cache but their approach is to improve future generic cache designs rather than improving specific multi-core applications. Klinkenberg et al. [19] describe H2M which is a heuristic tool for managing data placement in complex memory architectures in high performance computers (HPC, i.e. super computers). However, they use fixed hand made heuristics and are concerned only with runtime rather than seeking to show evolution can in general optimise last level cache (LLC) use by application software. They agree that managing diverse memory in parallel computing environments by hand is hard and yet will become increasingly important. Cloud White [40] is a tool for monitoring LLC contention between different customers' virtual machines (VMs) when they run on the same multi-core cloud computer server. Pons et al. [40] claim low overhead, but Cloud White is a black-box tool for monitoring Quality of Service (QoS) rather than an optimisation tool. Whereas Clite [55] uses

Bayesian Optimization to try to get the best mix of existing VMs rather than optimising individual applications.

3   Magpie

MAGPIE (Machine automated general performance improvement via evolution of software) [8] is a freely available genetic improvement system written in Python[1] and designed to be applied to software written in any programming language. The current release was downloaded from GitHub[2]. Magpie is well documented. For example, its GitHub pages include examples and tutorials. Also there is a more formal description [8]. Although we have used GI to optimise both code and parameters before [28], Magpie is unique in being a general genetic improvement framework that can optimise simultaneously parameters and any programming language. Parameters to be optimised might be, for example: constants[3], program command line and execution parameters (Sect. 4.2) and/or compiler options (Sect. 4.3). While much existing GI work has been based on lines of source code (which Magpie also supports), we use its ability to work with source code at the compiler's AST level by using XML trees.

4   PARSEC VIPS Thumbnail Benchmark

The VIPS image processing library was downloaded as part of PARSEC 3.0 from GitHub[4]. PARSEC as a whole is enormous, even the VIPS source library (sub directory `pkgs/apps/vips/···/src/libvips`) contains more than 90 000 lines of code (mostly C source code).

4.1   Profiling VIPS Thumbnail, Targeting C Code, Generating XML

As mentioned in the introduction, we chose the VIPS thumbnail benchmark from the VIPS library. vipsthumbnail.exe was compiled and linked following the VIPS installation documentation and profiled using the Linux perf profiling utility (perf version 3.10.0) operating at its maximum sampling frequency (40 000 Hz). perf collected data from ten runs with a variety of number of concurrent threads (–vips-concurrency). In all cases run time was dominated by the shrink_gen function. Remember VIPS is essentially a library, most of which is not used by an individual application. To extricate the important code used by vipsthumbnail.exe, we took the union of functions in the hierarchical call of shrink_gen and any function sampled by perf. This gave us 37 .c source files

---

[1] We use Python 3.10.1.

[2] https://github.com/bloa/magpie (last update before submission 2 October 2023).

[3] Our work evolving 50 000 parameters for RNAfold's free energy minimisation algorithm [37] and evolving 512 floating point values to convert the GNU C square root function into other functions [33], was done before Magpie was available.

[4] https://github.com/bamos/parsec-benchmark Version 3.0 for 64-bit x86.

containing 10 829 lines of C code. Notice this is not the whole of the VIPS thumbnail benchmark, it is still necessary to link to the libvips.so shared object library, but the 37 files do contain important code which we wish Magpie to optimise. A further filtering operation was done to select just the functions that are used during fitness testing (Sect. 5), reducing the 37 files to a total of 7 328 lines of code. These were converted to XML using scrml version 1.0.0 and made available to Magpie to tune.

4.2 VIPS Thumbnail Parameters

The VIPS thumbnail benchmark allows up to 12 command line parameters, however some of these change the output. Excluding these, left five (`vips-concurrency`, `vips-tile-width`, `vips-tile-height`, `vips-thinstrip-height` and `vips-fatstrip-height`) all of which were made available via a parameter file to Magpie to tune. Magpie starts its search from the default values.

4.3 GCC Compiler and Linker Options

The GNU compiler/linker version 10.2.1 has several hundred command line options. Rather than use them all, we selected those that appear in the installation scripts for VIPS plus some commonly used compilation options. For each, we set the default to the value used in the VIPS installation process but allowed Magpie the full range of allowed values. For example, VIPS uses -O2, so by default Magpie uses -O2. Although -O3 is available to Magpie, it was not used in the successful patches (Sect. 6). GCC options available to Magpie to tune are: `-fPIC -O -DNDEBUG -fvisibility -std=c99 -msse4.1 -fno-exceptions -ffat-lto-objects -flto -fno-strict-aliasing -fopenmp -fstack-protector -fstack-protector-strong -ftree-vectorize -g -m64 -mtune=generic -nostdlib --param=ssp-buffer-size -pipe -std=c++11 -std=c++98`

4.4 Magpie Local Search Parameters

Due to the noisy nature of LLC cache usage, Magpie was run 100 times but each run was allowed only 100 local search steps. Otherwise the Magpie defaults, such as default time out for fitness evaluation (30 s) and limit on output generated during fitness testing (10 000 bytes), were used.

Almost the full suite of Magpie's XML mutation operators were enabled: literal numbers, StmtReplacement, StmtInsertion, StmtDeletion, ComparisonOperatorSetting, ArithmeticOperatorSetting, NumericSetting, RelativeNumericSetting.

Magpie keeps track of the C source code it has mutated (via XML) and so only the mutated C code needs to be recompiled. In contrast, if the compiler command line is changed, all 37 C source files must be recompiled.

The mutated vipsthumbnail.exe was run with a command line generated by Magpie from the five variable VIPS thumbnail command line parameters (see Sect. 4.2).

5   Fitness Function

The experiments were run on a standard networked Centos 7 desktop. Even when apparently idle, it has more than two hundred active Linux processes, all of which use the LLC cache. LLC cache measurements are noisy (see, for example, Fig. 4. We used the Linux perf tool's API to measure cache usage during the critical multi-threaded image processing operations which create the thumbnail[5]. This allows us to isolate it from mundane operations, like processing the command line and reading and writing the image files. However initial experiments to reduce noise by placing the LLC cache in a defined state before starting perf measurements were unsuccessful. Instead the unmutated code was used as a reference and fitness is based on running it and then running the mutant as soon as possible, then setting fitness to the (signed) difference between their cache usage. Unfortunately even this paired approach is still quite noisy. (LLC measurements for the original unmutated code show that the coefficient of variation is 14%.) Indeed Sect. 6 suggests Fig. 4) shows running original and mutant as a pair failed to eliminate fitness cache measurement noise.

To summarise there are multiple aspects of a mutation's fitness: 1) If Magpie mutated one or more XML files (Sects. 4.1 and 4.4) or it changed the GCC options (Sect. 4.3), do the C source files still compile and link without error, 2) Does the mutant program run ok with the possibly mutated command line (Sect. 4.2), 3) Does it, within the two second timeout, produce an output file[6], 4) Is the output identical, 5) Finally, fitness is the number of LLC cache accesses by the reference program minus that of the mutant (remember Magpie minimises fitness). If any of the tests 1)–4) fail, Magpie discards the mutation.

6   Results

Magpie was run one hundred times on an otherwise idle 8 core 3.6 GHz Intel i7-4790 networked desktop with an 8 Mbyte LLC (L3) cache. Each time Magpie was allowed up to one hundred search steps (see Fig. 3). The whole 100 runs took less than five hours, cf. horizontal axis in Fig. 5. Mean Magpie run time 2:47 min:seconds each.

Figure 3 plots the training fitness of patches during each run. 38 Magpie runs terminated early and did not produce a best of run patch (black o). Figure 3 splits the remaining 62 Magpie runs into 21+ whose best of run patch failed to

---

[5] There is a vipsthumbnail command line option to allow the user to control the number of threads used during thumbnail image creation: –vips-concurrency, Sect. 4.2.

[6] The Linux `limit filesize` command can be used to restrict the total size of files generated but this was not necessary in these experiments.

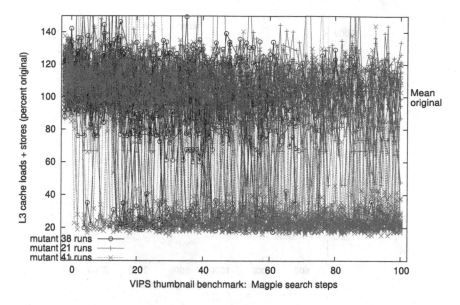

Fig. 3. 100 runs minimising last level cache accesses (load + stores) during multiple threaded processing reducing 3264 × 2448 image (23 970 833 bytes) to 128 × 96 (Fig. 2). 38 runs did not complete ○, 21 produced poor patches +, 41 reduced LLC cache usage 4.0 fold ×. See also Figs. 4 and 5.

generalise and 41× where it did (see also Figs. 4 and 5). Figure 3 shows by half way through, runs whose best patch will generalise are doing better (remember we are minimising) than the others, whose fitness tends to be scattered about the mean performance of the original C code (cf. 100% on vertical axis in Fig. 3). (In an effort to deal with noise, Magpie by default, performs three warmup fitness evaluations before commencing its search, hence in Fig. 3 the horizontal axis starts at -2 rather than 1.)

To counter over fitting and measure out-of-sample performance, the 62 best of run patches each were individually tested again 100 times. As with fitness testing (Sect. 5), each test consists of executing the reference unmutated code and the mutant as a pair, measuring their cache usage and checking the mutant still produces identical output. Figure 4 shows the cache use of the original code (x-axis) and of the mutant (y-axis). Notice noisy scatter of the data. Figure 4 (blue dots top) shows 21 mutants failed to give an improvement when run again. Their individual performance is pretty close to that of the original code (cf. the diagonal line in Fig. 4). Indeed the noise is also similar, giving rise to the approximately circular pattern at the top of Fig. 4). Also note the circular pattern indicates little correlation between the two measurements, suggesting the pairing of the reference and mutated code is ineffective at noise suppression in this case. Which hints that any pattern (if any) in the noise takes place faster than the <30 milliseconds between running the reference and mutated programs. In contrast 41 patches, despite the noise, always give better performance. (Plotted as red dots

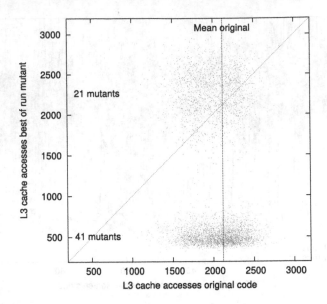

Fig. 4. LLC cache use of 62 best of run mutants, each tested 100 times (vertical axis). Horizontal axis LLC cache use of original code. Blue dots (near diagonal line) show 21 mutants which fail to improve on original code. Red dots (lower) show 41 mutants are always better than the unmutated code. Also Figure 5

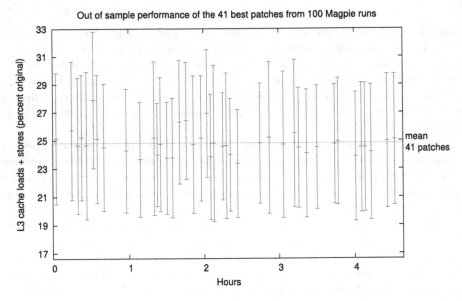

Fig. 5. 100 Magpie runs. Mean + and estimated standard deviation (error bars) of last level cache (LLC) use of the 41 good best of run patches found by Magpie (average of one hundred samples, cf. Figure 4). The horizontal axis shows when each patch was reported. The first successful mutant (left most) was found by Magpie after 2:42 minutes:seconds.

in the lower part of Fig. 4.) Although each of the 41 patches is in detail different, their performance are remarkably similar and each gives a ≈4.0× reduction in LLC cache usage, Fig. 5.

7   Follow up Experiment: Optimising VIPS Parameters

7.1   Types of Improvement Found in 41 Successful Magpie Runs

In the best of run mutants in the 41 runs which produced good patches there were between 1 and 10 individual changes (mean 4.9, total 201). There are only 15 GCC command line changes, none of which seem related to optimisation. For example, -O is not used. This may be because the available compiler and linker options aim to reduce the time taken for computation rather than data access. In contrast, all but two of the 41 patches tune one or more VIPS application parameters. Similarly, all but one of these 41 mutations changes one or more C source code files. In total there are: 8 C code insertions, 9 replacements and 23 statement deletions. Of the "smaller" code mutations, there are 20 arithmetic operator and 24 comparison changes, 27 direct changes to numbers and 30 relative changes (e.g. increasing by 50%). And 60 GCC or VIPS parameter changes. It is difficult to evaluate all the code changes but some appear not to matter as, despite removing whole unused functions before running Magpie (Sect. 4.1) they change code that is not executed or make a syntax change but the code semantics are unchanged, e.g. replace 0 by (0/2). Table 1 further summarises the 201 genes from the successful patches by mutation type and C source file.

7.2   Optimising vips-thinstrip-height and vips-fatstrip-height

From the top of Table 1 it is clear that VIPS parameter tuning stands out amongst the changes in the 41 successful Magpie runs. Therefore a second set of Magpie experiments were run to optimise LLC cache use by tuning only the two VIPS application parameters which occurred in almost all of the 41 successful mutants found in Sect. 6. Magpie was set up identically except: the GCC command line arguments and XML files and XML mutations were not used. Only vips-thinstrip-height and vips-fatstrip-height where included in the Magpie VIPS parameters. As before Magpie started from their default values (1 and 16 respectively) and again for both Magpie chose integer mutation values uniformly at random from the range 1 to 1200. Since it was no longer necessary to compile and link each mutant, Magpie runs were much faster (18.5 s).

In these final Magpie runs, the search space is greatly reduced, from effectively infinite to $[1200]^2$ (1 440 000). With 100 runs, each with up to 100 samples, in total Magpie was able to approximately sample the search space (see Fig. 6). The combined sampling suggests that vips-thinstrip-height correlates well with LLC cache usage. Therefore Fig. 7 concentrates upon it. Figure 6 shows values above vips-thinstrip-height = 6 simply scatter about the mean[7] and so they

---

[7] Although Table 2 suggests a slight downward trend in perf last level cache LLC measurements with increasing vips-thinstrip-height, this is not visible in Fig. 6, which includes both vips-thinstrip-height and vips-fatstrip-height.

Table 1. Distribution of 201 genes in the 41 best of run Magpie patches. The table is sorted by frequency, most important first, and is read left to right in row order. Thus VIPS thumbnail command line parameters and GCC compilation switches are mixed with source code arithmetic constants and XML changes. The numbers give each gene's frequency.

Type	parameter name/file		Type	parameter name/file	
VIPS	vips-thinstrip-height	39	Arith XML	im_shrink.c.xml	10
Arith XML	im_guess_prefix.c.xml	7	Arith XML	threadpool.c.xml	6
Arith XML	rw_mask.c.xml	5	Arith XML	memory.c.xml	5
Arith XML	im_vips2ppm.c.xml	5	Arith XML	im_prepare.c.xml	5
Arith XML	im_demand_hint.c.xml	5	Arith XML	format.c.xml	5
XML	im_demand_hint.c.xml	4	XML	check.c.xml	4
Arith XML	time.c.xml	4	Arith XML	im_embed.c.xml	4
Arith XML	check.c.xml	4	Arith XML	buffer.c.xml	4
XML	window.c.xml	3	XML	meta.c.xml	3
XML	im_init_world.c.xml	3	VIPS	vips-tile-width	3
VIPS	vips-fatstrip-height	3	GCC	-fstack-protector	3
Arith XML	object.c.xml	3	Arith XML	init.c.xml	3
Arith XML	im_conv.c.xml	3	Arith XML	im_affine.c.xml	3
Arith XML	debug.c.xml	3	XML	time.c.xml	2
XML	region.c.xml	2	XML	interpolate.c.xml	2
XML	init.c.xml	2	XML	im_guess_prefix.c.xml	2
XML	debug.c.xml	2	GCC	-mtune=generic	2
GCC	-ftree-vectorize	2	GCC	-fopenmp	2
Arith XML	window.c.xml	2	Arith XML	util.c.xml	2
Arith XML	region.c.xml	2	Arith XML	rect.c.xml	2
Arith XML	im_open.c.xml	2	Arith XML	im_copy.c.xml	2
Arith XML	im_close.c.xml	2	XML	util.c.xml	1
XML	semaphore.c.xml	1	XML	object.c.xml	1
XML	memory.c.xml	1	XML	im_vips2ppm.c.xml	1
XML	im_prepare.c.xml	1	XML	im_copy.c.xml	1
XML	im_convsep_f.c.xml	1	XML	im_convsep.c.xml	1
XML	im_affine.c.xml	1	XML	buffer.c.xml	1
GCC	-m64	1	GCC	-g	1
GCC	-fvisibility	1	GCC	-fstack-protector-strong	1
GCC	-fno-strict-aliasing	1	GCC	-flto	1
Arith XML	sinkdisc.c.xml	1	Arith XML	semaphore.c.xml	1
Arith XML	interpolate.c.xml	1			

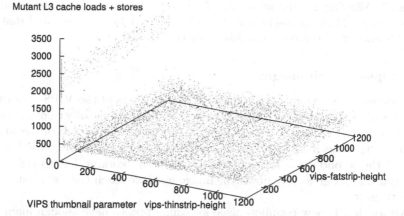

Fig. 6. 2nd Magpie experiment. LLC cache measurements during 100 runs to simultaneously tune vips-thinstrip-height and vips-fatstrip-height. Data from the same run have the same colour. The vertical line of dots above (1,16) are the initial default starting point for all 100 runs.

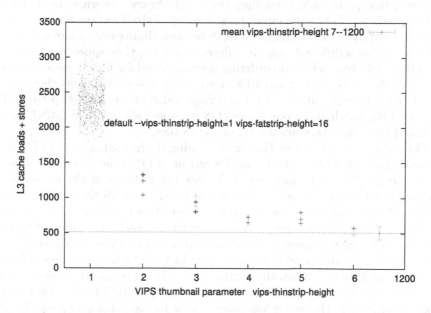

Fig. 7. 2nd Magpie experiment. LLC cache measurements during 100 runs to simultaneously tune vips-thinstrip-height and vips-fatstrip-height (the same data as Figure 6). Values above vips-thinstrip-height=6 not plotted as data are simply scattered about the mean (dotted line). Horizontal noise added to spread data for vips-thinstrip-height=1.

are omitted from Fig. 7). Instead their mean is plotted with a horizontal line in Fig. 7. Allowing for the noise, Fig. 7 suggests a monotonic reduction in LLC usage as vips-thinstrip-height is increased from its default value 1 and that vips-thinstrip-height ≥ 6 gives a 4.0 fold reduction in LLC use.

7.3   vips-thinstrip-height

This section tries to explain Magpie's results in terms of the VIPS application.

Two dimensional images, such as photographs, are usually laid out on disk and in memory as rows of consecutive pixels. Starting at the left of the top edge and moving along it to the right edge and then moving down to the left hand edge of the second row. This pattern is repeated, working left to right across each row and progressively down the image until we reach the right hand end of its bottom row.

Although it is now common place for computers to have enough main RAM memory to store uncompressed the whole image, often images are too big to fit into the cache. For example, excluding metadata, a full colour P6 (3 bytes per pixel) 3264×2448 image occupies $3 \times 3264 \times 2448 = 23\,970\,816$ bytes, whereas these experiments were run on a computer with a LLC cache of $8\,388\,608$ bytes. At more than 7000 lines of deeply nested [29] multi-threaded code, it is difficulty to be exactly sure which actions impact the LLC cache and in which ways. However, to exploit multi-threading, the VIPS library (ignoring small overlaps and edge effects) divides the input image into equal sized rectangular tiles. These are processed by separate threads and so random timing effects mean that they are processed in a different order in different runs, but they approximately follow the left to right top to bottom ordering normally used for image processing. The tiles are 128 pixels (384 bytes) wide. Depending upon alignment, they occupy 6 or 7, 64 byte cache lines. With the default value of vips-thinstrip-height (1), they are 1 pixel high, so for our 3264×2448 example each row takes $\lceil 3264/128 \rceil =$ 26 (25.5) tiles, and there are a total of 63 648 tiles.

On average, see Fig. 4 and Table 2, processing the image takes 2161 LLC cache accesses. 1909 are LLC cache loads, the rest are LLC cache stores. Meaning on average each LLC cache load access fetches 12 600 bytes of the image. What appears to be happening is the LLC cache is asked for 26 tiles of data. These occupy 9 9792 bytes (depending upon alignment, this is 153 or 155 cache lines). Even though these data requests arrive at different times from different threads, the cache hierarchy appears to be able to consolidate these into a single LLC access. (9 9792 is within 28% of the average LLC cache load size. See column 9 in top row of Table 2.) When these data arrive, the 8 threads are able to process the 26 tiles and then request data for the next row (again 153 or 155 cache lines). It is not clear why the cache hierarchy is able to consolidate requests for 25.5 tiles but not for multiple rows.

When vips-thinstrip-height is increased to 2, the VIPS tiles are increased from 128×1 to 128×2 pixels (again ignoring overlaps). Again depending upon alignment, each tile now occupies 12 or 14 cache lines. and the number of LLC cache read accesses falls on average to 924, i.e. 25 900 bytes each. This roughly

Table 2. Mean impact on last level cache LLC of running the original code using 8 threads with various values of –vips-thinstrip-height (column 1) 100 times. Numbers in brackets are estimates of standard deviation, except last column where () indicates standard error in the calculation of $\frac{\text{mean LLC load size}}{\text{rows size}}$ (column 9 = column 6 divided by column 8.)

thinstrip	LCC loads		stores		load bytes		rows bytes	Ratio	
1	1909	(290)	252	(22)	12600	(1910)	9792	1.28	(0.020)
2	924	(191)	188	(19)	25900	(5340)	19584	1.32	(0.027)
3	764	(113)	173	(13)	31400	(4650)	29376	1.07	(0.016)
4	609	(108)	166	(13)	39300	(7010)	39168	1.00	(0.018)
5	578	(89)	165	(13)	41500	(6360)	48960	0.85	(0.013)
6	477	(92)	155	(12)	50200	(9690)	58752	0.85	(0.017)
7	450	(84)	152	(12)	53200	(9870)	68544	0.78	(0.014)
8	447	(82)	152	(12)	53700	(9830)	78336	0.69	(0.013)
9	455	(74)	151	(12)	52700	(8540)	88128	0.60	(0.010)
10	432	(80)	147	(12)	55500	(10230)	97920	0.57	(0.010)
20	424	(93)	149	(14)	56600	(12450)	195840	0.29	(0.006)
40	410	(81)	145	(11)	58500	(11580)	391680	0.15	(0.003)
60	371	(69)	145	(12)	64500	(11950)	587520	0.11	(0.002)
80	383	(75)	148	(12)	62600	(12250)	783360	0.08	(0.002)
96	341	(62)	145	(11)	70200	(12840)	940032	0.07	(0.001)
100	353	(72)	146	(11)	67800	(13830)	979200	0.07	(0.001)
500	350	(84)	146	(12)	68500	(16370)	4896000	0.01	(0.000)
1000	345	(61)	146	(11)	69400	(12340)	9792000	0.01	(0.000)
1500	345	(57)	145	(12)	69400	(11450)	14688000	0.00	(0.000)
2000	364	(75)	144	(10)	65900	(13580)	19584000	0.00	(0.000)
2448	342	(57)	146	(10)	70000	(11560)	23970816	0.00	(0.000)
2500	347	(64)	146	(11)	69000	(12640)	23970816	0.00	(0.000)

doubling of the mean size of data per LLC cache load is consistent with the idea that the cache hierarchy is able to consolidate 26 scattered request for cache reads but something about reading the end of two rows of consecutive bytes prevents further consolidation.

Table 2 shows as vips-thinstrip-height is increased to 3, 4, 5 and 6, the average number of bytes fetched per LLC load access increases and remains approximately the same as the size of one row of tiles. I.e., the last but one column in Table 2 remain near 1.0.

Although vips-thinstrip-height can be increased above 6, from Table 2 it is clear that this has only a small effect. It may be the VIPS code itself places a limit on the tile height. Alternatively the computer's RAM and/or the cache hierarchy may limit pre-fetch sizes to 64 Kbytes (65 536 bytes), cf. column 6 in

Table 2. Above vips-thinstrip-height=96 (a magic number for VIPS' tile height) there appear to be no further reductions.

It is clear that the interaction between vips-thinstrip-height and the LLC is noisy and complicated but its interaction with internal VIPS tile height seems like a reasonable first step at explaining how it works and why vips-thinstrip-height stands out in Magpie's "hands clean" optimisation.

8 Conclusions

We have shown an "off the shelf" genetic improvement (GI) system Magpie [8] using multiple hill climbing runs can automatically tune a standard parallel processing benchmark to reduce four fold its use of last level cache (LLC) on modern multi-core computers.

The VIPS image processing thumbnail benchmark is representative of a large class of parallel processing programs. Its multi-processing is based on POSIX pthreads, which is heavily used in multi-core applications. Although we expect continued growth of hybrid computers which off load significant computation to accelerators (e.g. GPU [46], TPUs [18], FPGAs [50]), we expect they will remain hard to program effectively and so they may remain the preserve of artificial intelligence (AI) deep artificial neural networks [53], e.g. for training large language models (LLMs) and specific domains such as astronomy [1] and bioinformatics [49]. Instead we anticipate mundane applications will continue to be run on multi-core computers. To get the best of their increasing numbers of cores will require, not just optimising the code, but also increasingly optimising data communication. However it appears (cf. Sect. 7.1) existing compilers optimise computation not data access ("Compilers are not good at managing caches" [4, p44]). Therefore new tools will be needed [58] to make effective use of complex proprietary hardware cache hierarchies, whose operation is invisible to the user level programmer. By taking a "hands off" approach Magpie may be able to help programmers by optimising for them the last level cache which provides high bandwidth inter-core communication, which will be increasingly needed in what promises to be the dominant domain for future software development.

Acknowledgements. I am grateful for the help of Aymeric Blot and Dan Blackwell.

References

1. Adamek, K., Dimoudi, S., Giles, M., Armour, W.: GPU fast convolution via the overlap-and-save method in shared memory. ACM Trans. Archit. Code Optim. **17**(3), article no 18 (2020). https://doi.org/10.1145/3394116
2. Andrews, R.J., et al.: A map of the SARS-CoV-2 RNA structurome. NAR Genomics Bioinform. **3**(2), lqab043 (2021). https://doi.org/10.1093/nargab/lqab043
3. Banzhaf, W., Nordin, P., Keller, R.E., Francone, F.D.: Genetic Programming - An Introduction. Morgan Kaufmann, London (1998). https://www.amazon.co.uk/Genetic-Programming-Introduction-Artificial-Intelligence/dp/155860510X

4. Berger, M.: Compilers and computer architecture: caches and caching. G5035, BSc/MComp Computer Science, University of Sussex, December 2019. https://users.sussex.ac.uk/mfb21/compilers/slides/15-handout.pdf. Accessed November 2023

5. Bienia, C., Kumar, S., Singh, J.P., Li, K.: The PARSEC benchmark suite: characterization and architectural implications. In: Moshovos, A., Tarditi, D., Olukotun, K. (eds.) 17th International Conference on Parallel Architectures and Compilation Techniques, PACT 2008, pp. 72–81. ACM, Toronto, Ontario, Canada, 25–29 October 2008. https://doi.org/10.1145/1454115.1454128

6. Blot, A., Petke, J.: Comparing genetic programming approaches for non-functional genetic improvement. In: Hu, T., Lourenço, N., Medvet, E., Divina, F. (eds.) EuroGP 2020. LNCS, vol. 12101, pp. 68–83. Springer, Cham (2020). https://doi.org/10.1007/978-3-030-44094-7_5

7. Blot, A., Petke, J.: Empirical comparison of search heuristics for genetic improvement of software. IEEE Trans. Evol. Comput. 25(5), 1001–1011 (2021). https://doi.org/10.1109/TEVC.2021.3070271

8. Blot, A., Petke, J.: MAGPIE: machine automated general performance improvement via evolution of software, 4 August 2022. arXiv. http://dx.doi.org/10.48550/arxiv.2208.02811

9. Blot, A., Petke, J.: Using genetic improvement to optimise optimisation algorithm implementations. In: Hadj-Hamou, K. (ed.) 23ème congrès annuel de la Société Française de Recherche Opérationnelle et d'Aide à la Décision, ROADEF'2022. INSA Lyon, France, 23–25 February 2022. https://hal.archives-ouvertes.fr/hal-03595447

10. Brownlee, A.E.I., et al.: Enhancing genetic improvement mutations using large language models. In: Arcaini, P., Tao Yue, Fredericks, E. (eds.) Search-Based Software Engineering. SSBSE 2023: Challenge Track. LNCS, vol. 14415, pp. 153–159. Springer, Cham (2023). https://doi.org/10.1007/978-3-031-48796-5_13

11. Bruce, B.R.: Automatically exploring computer system design spaces. In: Bruce, B.R., et al. (eds.) GI @ GECCO 2022, pp. 1926–1927. Association for Computing Machinery, Boston, USA, 9 July 2022. https://doi.org/10.1145/3520304.3534021

12. de Almeida Farzat, F., de Oliveira Barros, M., Horta Travassos, G.: Challenges on applying genetic improvement in JavaScript using a high-performance computer. J. Softw. Eng. Res. Dev. 6(12) (2018). https://doi.org/10.1186/s40411-018-0056-2, 20th Iberoamerican Conference on Software Engineering

13. Griffin, D., Stepney, S., Vidamour, I.: DebugNS: novelty search for finding bugs in simulators. In: Nowack, V., et al. (eds.) 12th International Workshop on Genetic Improvement @ICSE 2023, pp. 17–18. IEEE, Melbourne, Australia, 20 May 2023. https://doi.org/10.1109/GI59320.2023.00012

14. Harman, M., Jones, B.F.: Search based software engineering. Inf. Softw. Technol. 43(14), 833–839 (2001). https://doi.org/10.1016/S0950-5849(01)00189-6

15. Liou, J.-Y., Forrest, S., Wu, C.-J.: Genetic improvement of GPU code. In: Petke, J., Tan, S.H., Langdon, W.B., Weimer, W. (eds.) GI-2019, ICSE Workshops Proceedings, pp. 20–27. IEEE, Montreal, 28 May 2019. https://doi.org/10.1109/GI.2019.00014, Best Paper

16. Liou, J.-Y., Wang, X., Forrest, S., Wu, C.-J.: GEVO: GPU code optimization using evolutionary computation. ACM Trans. Archit. Code Optim. 17(4), Article 33 (2020). https://doi.org/10.1145/3418055

17. Jimenez, D.A., Teran, E., Gratz, P.V.: Last-level cache insertion and promotion policy in the presence of aggressive prefetching. IEEE Comput. Archit. Lett. 22(1), 17–20 (2023). https://doi.org/10.1109/LCA.2023.3242178

18. Jouppi, N.P., et al.: TPU v4: an optically reconfigurable supercomputer for machine learning with hardware support for embeddings. In: Proceedings of the 50th Annual International Symposium on Computer Architecture, ISCA, p. article no 82. ACM, Orlando, FL, USA (2023). https://doi.org/10.1145/3579371.3589350

19. Klinkenberg, J., et al.: H2M: exploiting heterogeneous shared memory architectures. Futur. Gener. Comput. Syst. **148**, 39–55 (2023). https://doi.org/10.1016/J.FUTURE.2023.05.019

20. Klus, P., et al.: BarraCUDA - a fast short read sequence aligner using graphics processing units. BMC Res. Notes **5**(27) (2012). https://doi.org/10.1186/1756-0500-5-27

21. Koza, J.R.: Genetic Programming: On the Programming of Computers by Means of Natural Selection. MIT Press, Cambridge, MA, USA (1992). http://mitpress.mit.edu/books/genetic-programming

22. Krauss, O.: Exploring the use of natural language processing techniques for enhancing genetic improvement. In: Nowack, V., et al. (eds.) 12th International Workshop on Genetic Improvement @ICSE 2023, pp. 21–22. IEEE, Melbourne, Australia, 20 May 2023. https://doi.org/10.1109/GI59320.2023.00014

23. Kuepper, J., et al.: CryptOpt: verified compilation with randomized program search for cryptographic primitives. In: Foster, N. (ed.) 44th ACM SIGPLAN Conference on Programming Language Design and Implementation, PLDI 2023, p. article no. 158. Association for Computing Machinery, Orlando, Florida, 17–21 June 2023. https://doi.org/10.1145/3591272, Gold winner 2023 HUMIES, PLDI Distinguished Paper

24. Langdon, W.B., Lam, B.Y.H.: Genetically improved BarraCUDA. BioData Min. **20**(28) (2017). https://doi.org/10.1186/s13040-017-0149-1

25. Langdon, W.B., Harman, M.: Evolving a CUDA kernel from an nVidia template. In: Sobrevilla, P. (ed.) 2010 IEEE World Congress on Computational Intelligence, pp. 2376–2383. IEEE, Barcelona, 18–23 July 2010. https://doi.org/10.1109/CEC.2010.5585922

26. Langdon, W.B., Al-Subaihin, A., Blot, A., Clark, D.: Genetic improvement of LLVM intermediate representation. In: Pappa, G., Giacobini, M., Vasicek, Z. (eds.) Genetic Programming. EuroGP 2023. LNCS, vol. 13986, pp. 244–259. Springer, Cham (2023). https://doi.org/10.1007/978-3-031-29573-7_16

27. Langdon, W.B., Alexander, B.J.: Genetic improvement of OLC and H3 with magpie. In: Nowack, V., et al. (eds.) 12th International Workshop on Genetic Improvement @ICSE 2023, pp. 9–16. IEEE, Melbourne, Australia, 20 May 2023. https://doi.org/10.1109/GI59320.2023.00011

28. Langdon, W.B., Brian Yee Hong Lam, Petke, J., Harman, M.: Improving CUDA DNA analysis software with genetic programming. In: Silva, S., et al. (eds.) GECCO '15: Proceedings of the 2015 Annual Conference on Genetic and Evolutionary Computation, pp. 1063–1070. ACM, Madrid, 11–15 July 2015. https://doi.org/10.1145/2739480.2754652

29. Langdon, W.B., Clark, D.: Deep mutations have little impact. In: Gabin, A., et al. (eds.) 13th International Workshop on Genetic Improvement @ICSE 2024. ACM, Lisbon, 16 April 2024, forthcoming

30. Langdon, W.B., Harman, M.: Genetically improved CUDA C++ software. In: Nicolau, M., et al. (eds.) EuroGP 2014. LNCS, vol. 8599, pp. 87–99. Springer, Heidelberg (2014). https://doi.org/10.1007/978-3-662-44303-3_8

31. Langdon, W.B., Harman, M.: Grow and graft a better CUDA pknotsRG for RNA pseudoknot free energy calculation. In: Langdon, W.B., Petke, J., White, D.R.

(eds.) Genetic Improvement 2015 Workshop, pp. 805–810. ACM, Madrid, 11–15 July 2015. https://doi.org/10.1145/2739482.2768418

32. Langdon, W.B., Harman, M.: Optimising existing software with genetic programming. IEEE Trans. Evol. Comput. **19**(1), 118–135 (2015). https://doi.org/10.1109/TEVC.2013.2281544

33. Langdon, W.B., Krauss, O.: Genetic improvement of data for maths functions. ACM Trans. Evolut. Learn. Optim. **1**(2), Article No.: 7 (2021). https://doi.org/10.1145/3461016

34. Langdon, W.B., Lorenz, R.: Evolving AVX512 parallel C code using GP. In: Sekanina, L., Hu, T., Lourenço, N., Richter, H., García-Sánchez, P. (eds.) EuroGP 2019. LNCS, vol. 11451, pp. 245–261. Springer, Cham (2019). https://doi.org/10.1007/978-3-030-16670-0_16

35. Langdon, W.B., Modat, M., Petke, J., Harman, M.: Improving 3D medical image registration CUDA software with genetic programming. In: Igel, C., et al. (eds.) GECCO '14: Proceeding of the Sixteenth Annual Conference on Genetic and Evolutionary Computation Conference, pp. 951–958. ACM, Vancouver, BC, Canada, 12–15 July 2014. https://doi.org/10.1145/2576768.2598244

36. Langdon, W.B., Petke, J., Blot, A., Clark, D.: Genetically improved software with fewer data caches misses. In: Silva, S., et al. (eds.) Proceedings of the Companion Conference on Genetic and Evolutionary Computation, pp. 799–802. GECCO '23, Association for Computing Machinery, Lisbon, Portugal, 15–19 July 2023. https://doi.org/10.1145/3583133.3590542

37. Langdon, W.B., Petke, J., Lorenz, R.: Evolving better RNAfold structure prediction. In: Castelli, M., Sekanina, L., Zhang, M., Cagnoni, S., García-Sánchez, P. (eds.) EuroGP 2018. LNCS, vol. 10781, pp. 220–236. Springer, Cham (2018). https://doi.org/10.1007/978-3-319-77553-1_14

38. Langdon, W.B., et al.: Genetic improvement of GPU software. Genet. Program. Evolvable Mach. **18**(1), 5–44 (2017). https://doi.org/10.1007/s10710-016-9273-9

39. Lorenz, R., et al.: ViennaRNA package 2.0. Algorithms Mol. Biol. **6**(1) (2011). https://doi.org/10.1186/1748-7188-6-26

40. Pons, L., et al.: Cloud White: Detecting and estimating QoS degradation of latency-critical workloads in the public cloud. Futur. Gener. Comput. Syst. **138**, 13–25 (2023). https://doi.org/10.1016/J.FUTURE.2022.08.012

41. Marginean, A.: Automated Software Transplantation. Ph.D. thesis, University College London, UK, 8 November 2021. https://discovery.ucl.ac.uk/id/eprint/10137954/1/Marginean_10137954_thesis_redacted.pdf, ACM SIGEVO Award for the best dissertation of the year

42. Marginean, A., Barr, E.T., Harman, M., Jia, Y.: Automated transplantation of call graph and layout features into Kate. In: Barros, M., Labiche, Y. (eds.) SSBSE 2015. LNCS, vol. 9275, pp. 262–268. Springer, Cham (2015). https://doi.org/10.1007/978-3-319-22183-0_21

43. Martinez, K., Cupitt, J.: VIPS - a highly tuned image processing software architecture. In: Proceedings of the 2005 International Conference on Image Processing, ICIP, pp. 574–577. IEEE, Genoa, Italy, 11–14 September 2005. https://doi.org/10.1109/ICIP.2005.1530120

44. Moore, G.E.: Cramming more components onto integrated circuits. Electronics **38**(8), 114–117 (1965)

45. Murphy, A., Laurent, T., Ventresque, A.: The case for grammatical evolution in test generation. In: Bruce, B.R., et al. (eds.) GI @ GECCO 2022, pp. 1946–1947. Association for Computing Machinery, Boston, USA, 9 July 2022. https://doi.org/10.1145/3520304.3534042

46. Owens, J.D., et al.: GPU computing. Proc. IEEE **96**(5), 879–899 (2008). Invited paper, https://doi.org/10.1109/JPROC.2008.917757

47. Petke, J., et al.: Genetic improvement of software: a comprehensive survey. IEEE Trans. Evolut. Comput. **22**(3), 415–432 (2018). https://doi.org/10.1109/TEVC.2017.2693219

48. Poli, R., Langdon, W.B., McPhee, N.F.: A field guide to genetic programming. Published via http://lulu.com and freely (2008). http://www.gp-field-guide.org.uk, (With contributions by J. R. Koza)

49. Robinson, T., Harkin, J., Shukla, P.: Hardware acceleration of genomics data analysis: challenges and opportunities. Bioinformatics **37**(13), 1785–1795 (2021). https://doi.org/10.1093/bioinformatics/btab017

50. Santiago, A., et al.: Analysis and deployment of applications acceleration environment for Xilinx hardware-accelerated platforms. In: 37th Conference on Design of Circuits and Integrated Circuits (DCIS). IEEE, Pamplona, Spain, 16–18 November 2022. https://doi.org/10.1109/DCIS55711.2022.9970101

51. Schweim, D., et al.: Using knowledge of human-generated code to bias the search in program synthesis with grammatical evolution. In: Chicano, F., et al. (eds.) Proceedings of the 2021 Genetic and Evolutionary Computation Conference Companion, pp. 331–332. GECCO '21, Association for Computing Machinery, internet, 10–14 July 2021. https://doi.org/10.1145/3449726.3459548

52. Shuyue Stella Li, et al.: Genetic improvement in the Shackleton framework for optimizing LLVM pass sequences. In: Bruce, B.R., et al. (eds.) GI @ GECCO 2022, pp. 1938–1939. Association for Computing Machinery, Boston, USA, 9 July 2022. https://doi.org/10.1145/3520304.3534000, winner Best Presentation

53. Silver, D., et al.: Mastering the game of Go without human knowledge. Nature **550**(7676), 354–359 (2017). https://doi.org/10.1038/nature24270

54. Kang, S., Yoo, S.: Towards objective-tailored genetic improvement through large language models. In: Nowack, V., et al. (eds.) 12th International Workshop on Genetic Improvement @ICSE 2023, pp. 19–20. IEEE, Melbourne, Australia, 20 May 2023. https://doi.org/10.1109/GI59320.2023.00013, Best position paper

55. Patel, T., Tiwari, D.: CLITE: efficient and QoS-aware co-location of multiple latency-critical jobs for warehouse scale computers. In: 2020 IEEE International Symposium on High Performance Computer Architecture (HPCA), pp. 193–206 (2020). https://doi.org/10.1109/HPCA47549.2020.00025

56. Weimer, W.: From deep learning to human judgments: lessons for genetic improvement. GI @ GECCO 2022, 9 July 2022. http://geneticimprovementofsoftware.com/slides/gi2022gecco/weimer-keynote-gi-gecco-22.pdf, Invited keynote

57. Zhang, Y., Huang, Y.: Leveraging fuzzy system to reduce uncertainty of decision making in software engineering automation. In: Bruce, B.R., et al. (eds.) GI @ GECCO 2022, pp. 1948–1949. Association for Computing Machinery, Boston, USA, 9 July 2022. https://doi.org/10.1145/3520304.3533991

58. Lin, Y.-C., Lee, J.-K., Bodin, F.: Guest editorial: special issue on embedded multicore applications and optimization. J. Signal Process. Syst. **91**(3–4), 217–218 (2019). https://doi.org/10.1007/S11265-018-1431-2

Author Index

© The Editor(s) (if applicable) and The Author(s), under exclusive license
to Springer Nature Switzerland AG 2024
M. Giacobini et al. (Eds.): EuroGP 2024, LNCS 14631, p. 227, 2024.
https://doi.org/10.1007/978-3-031-56957-9

Printed in the United States
by Baker & Taylor Publisher Services

Printed in the United States
by Baker & Taylor Publisher Services